DEVELOPMENTS IN SEDIMENTOLOGY 23

INFLUENCE OF ABYSSAL CIRCULATION ON SEDIMENTARY ACCUMULATIONS IN SPACE AND TIME

FURTHER TITLES IN THIS SERIES

1. *L.M.J.U. VAN STRAATEN, Editor*
DELTAIC AND SHALLOW MARINE DEPOSITS

2. *G.C. AMSTUTZ, Editor*
SEDIMENTOLOGY AND ORE GENESIS

3. *A.H. BOUMA* and *A. BROUWER, Editors*
TURBIDITES

4. *F.G. TICKELL*
THE TECHNIQUES OF SEDIMENTARY MINERALOGY

5. *J.C. INGLE Jr.*
THE MOVEMENT OF BEACH SAND

6. *L. VAN DER PLAS*
THE IDENTIFICATION OF DETRITAL FELDSPARS

7. *S. DZULYNSKI* and *E.K. WALTON*
SEDIMENTARY FEATURES OF FLYSCH AND GREYWACKES

8. *G. LARSEN* and *G.V. CHILINGAR, Editors*
DIAGENESIS IN SEDIMENTS

9. *G.V. CHILINGAR, H.J. BISSELL* and *R.W. FAIRBRIDGE, Editors*
CARBONATE ROCKS

10. *P. McL. D. DUFF, A. HALLAM* and *E.K. WALTON*
CYCLIC SEDIMENTATION

11. *C.C. REEVES Jr.*
INTRODUCTION TO PALEOLIMNOLOGY

12. *R.G.C. BATHURST*
CARBONATE SEDIMENTS AND THEIR DIAGENESIS

13 *A.A. MANTEN*
SILURIAN REEFS OF GOTLAND

14. *K.W. GLENNIE*
DESERT SEDIMENTARY ENVIRONMENTS

15. *C.E. WEAVER* and *L.D. POLLARD*
THE CHEMISTRY OF CLAY MINERALS

16. *H.H. RIEKE III* and *G.V. CHILINGARIAN*
COMPACTION OF ARGILLACEOUS SEDIMENTS

17. *M.D. PICARD* and *L.R. HIGH Jr.*
SEDIMENTARY STRUCTURES OF EPHEMERAL STREAMS

18. *G.V. CHILINGARIAN* and *K.H. WOLF*
COMPACTION OF COARSE-GRAINED SEDIMENTS

19. *W. SCHWARZACHER*
SEDIMENTATION MODELS AND QUANTITATIVE STRATIGRAPHY

20. *M.R. WALTER, Editor*
STROMATOLITES

21. *B. VELDE*
CLAYS AND CLAY MINERALS IN NATURAL AND SYNTHETIC SYSTEMS

22. *C.E. WEAVER and K.C. BECK*
MIOCENE OF THE SOUTHEASTERN UNITED STATES

DEVELOPMENTS IN SEDIMENTOLOGY 23

INFLUENCE OF ABYSSAL CIRCULATION ON SEDIMENTARY ACCUMULATIONS IN SPACE AND TIME

EDITED BY

BRUCE C. HEEZEN

Lamont-Doherty Geological Observatory of Columbia University, Palisades, N.Y. (U.S.A.)

Reprinted from Marine Geology Vol. 23 No. 1/2

ELSEVIER SCIENTIFIC PUBLISHING COMPANY
AMSTERDAM – OXFORD – NEW YORK 1977

ELSEVIER SCIENTIFIC PUBLISHING COMPANY
335 Jan van Galenstraat
P.O. Box 211, Amsterdam, The Netherlands

Distributors for the United States and Canada:

ELSEVIER/NORTH-HOLLAND INC.
52, Vanderbilt Avenue
New York, N.Y. 10017

ISBN: 0444-41569-6

Printed in The Netherlands

PREFACE

The papers included in this special issue were given at a symposium entitled: "Influence of Abyssal Circulation on Sedimentary Accumulations in Space and Time" held August 27, 1975 in Grenoble, France during the General Assembly of the International Association for the Physical Sciences of the Ocean and the XVIth General Assembly of the International Union of Geodesy and Geophysics. The symposium, sponsored by the IAPSO Commission on Marine Geophysics was well attended and the discussions were spirited and informative.

The nature and total thickness of sediment lying on oceanic basement in a given area is largely determined by:

(1) the date of commencement of sedimentation;
(2) the initial depth (below sea level) of the juvenile crust;
(3) the history of productivity of the overlying waters;
(4) the presence of additional nonpelagic sources;
(5) the presence of processes of sediment redistribution.

The date of commencement can be estimated from magnetic anomaly stripes. These ages have been calibrated by Deep Sea Drilling Project holes to oceanic basement. The initial depth of the juvenile crust together with the original and subsequent levels of the calcium carbonate compensation depth in respect to the depositional surface determine the proportion of ooze and clay. The history of productivity in a given area may include not only initial ooze deposition followed by abyssal clay deposition on a subsiding crust but also a crossing or recrossing of the equatorial or polar front productivity belts with associated alternations of ooze and clay and episodes of higher and lower than normal rates of deposition. Up-wind injection of sediments carried into the atmosphere from volcanoes or desert areas can introduce significant variations in sedimentation. Turbidity currents can have an overwhelming influence in the areas they enter.

The first four factors all result in more or less sediment accumulation. The last one can also result in the removal and redistribution on the sea floor of previously deposited sediment. It is this later aspect which is the central theme of the papers presented in the present volume.

In the following papers the effects of abyssal circulation on sedimentation which are treated vary in magnitude from slight increases in suspended concentrations in sea water (Biscaye and Eittreim) to dissolution of planktonic tests (Johnson et al., Mallet and Heezen) to scour of sediments from beds of manganese nodules (Watkins and Kennett) to gentle scour observed from submersibles (Heezen and Rawson) to sharp-crested ripples and scour marks photographed (Stanley and Taylor) on a seamount to huge sand waves and scour channels revealed on deep-towed vehicle side-scan records and photographs (Lonsdale and Spiess) to the creation of isthmian barriers (Holcombe and Moore) and the stagnation of entire ocean basins (Ryan and Cita).

This convener wishes to thank both the speakers and the audience for their contribution to this successful symposium. We also thank Dr. Eugene LaFond, Secretary to IAPSO, and Professor Henry Lacombe, President of IAPSO, for their invaluable assistance.

BRUCE C. HEEZEN
(President, IAPSO Commission on Marine Geophysics)

CONTENTS

Marine Geology, 23 (1977) 1—33

VEMA CHANNEL PALEO-OCEANOGRAPHY: PLEISTOCENE DISSOLUTION CYCLES AND EPISODIC BOTTOM WATER FLOW*

DAVID A. JOHNSON[1], MICHAEL LEDBETTER[2] and LLOYD H. BURCKLE[3]

[1] *Woods Hole Oceanographic Institution, Woods Hole, Mass. (U.S.A.)*
[2]*Graduate School of Oceanography, University of Rhode Island, Kingston, R.I. (U.S.A.)*
[3]*Lamont-Doherty Geological Observatory, Palisades, N.Y. (U.S.A.)*

(Received April 28, 1976)

ABSTRACT

Johnson, D.A., Ledbetter, M. and Burckle, L.H., 1977. Vema Channel paleo-oceanography: Pleistocene dissolution cycles and episodic bottom water flow. Mar. Geol., 23: 1—33.

Investigations of piston cores from the Vema Channel and lower flanks of the Rio Grande Rise suggest the presence of episodic flow of deep and bottom water during the Late Pleistocene. Cores from below the present-day foraminiferal lysocline (at ~4000 m) contain an incomplete depositional record consisting of Mn nodules and encrustations, hemipelagic clay, displaced high-latitude diatoms, and poorly preserved heterogeneous microfossil assemblages. Cores from the depth range between 2900 m and 4000 m contain an essentially complete Late Pleistocene record, and consist of well-defined carbonate dissolution cycles with periodicities of ~100,000 years. Low carbonate content and increased dissolution correspond to glacial episodes, as interpreted by oxygen isotopic analysis of bulk foraminiferal assemblages. The absence of diagnostic high-latitude indicators (Antarctic diatoms) within the dissolution cycles, however, suggests that AABW may *not* have extended to significantly shallower elevations on the lower flanks of the Rio Grande Rise during the Late Pleistocene. Therefore episodic AABW flow may not necessarily be the mechanism responsible for producing these cyclic events. This interpretation is also supported by the presence of an apparently complete Brunhes depositional record in the same cores, suggesting current velocities insufficient for significant erosion. Fluctuations in the properties and flow characteristics of another water mass, such as NADW, may be involved.

The geologic evidence in core-top samples near the present-day AABW/NADW transition zone is consistent with either of two possible interpretations of the upper limit of AABW on the east flank of the channel. The foraminiferal lysocline, at ~ 4000 m, is near the top of the benthic thermocline and nepheloid layer, and may therefore correspond to the upper limit of relatively corrosive AABW. On the other hand, the carbonate compensation depth (CDD) at ~4250 m, which corresponds to the maximum gradient in the benthic thermocline, is characterized by rapid deposition of relatively fine-grained sediment. Such a zone of convergence and preferential sediment accumulation would be expected near the level of no motion in the AABW/NADW transition zone as a consequence of Ekman-layer veering of the mean velocity vector in the bottom boundary layer. It is possible that both of these interpretations are in part correct. The "level of no motion" may in fact correspond to the CCD, while at the same time relatively corrosive

*Contribution No. 3734 of the Woods Hole Oceanographic Institution.
Contribution No. 2365 of the Lamont-Doherty Geological Observatory.

water of Antarctic origin may mix with overlying NADW and therefore elevate the foraminiferal lysocline to depths above the level of no motion. Closely spaced observations of the hydrography and flow characteristics within the benthic thermocline will be required in order to use sediment parameters as more precise indicators of paleocirculation.

INTRODUCTION

In recent years, the erosion and redeposition of sediment on the deep ocean floor has come to be recognised as a process which is important over widespread geographic regions and throughout much of the geologic past. Observations of differential sediment accumulation patterns in seismic reflection profiling records, and the common occurrence of unconformities in sediment cores and at DSDP sites, indicate that the depositional record is generally incomplete. Numerous investigators have interpreted these deep-ocean unconformities as indicative of episodic bottom current flow, which in turn may result from changes in tectonic configurations and/or climatic conditions (e.g., Jones et al., 1970; Watkins and Kennett, 1971; Johnson, 1972; Berggren and Hollister, 1974). Particular attention has been focused on interpreting paleocirculation patterns during the Late Tertiary and Quaternary, inasmuch as major tectonic readjustments during this brief span of time have been minimal, and consequently changing patterns of sedimentation may be principally climatically controlled.

The various components of marine sediments represent a multiple of source regions, and may therefore reflect climatic conditions in both terrestrial, coastal, pelagic, and bathyal environments. Since most of the sediment components in abyssal regions originate primarily at distant terrestrial sources or in the photic zone in the overlying water, most studies of the Pleistocene marine record have been directed toward interpreting past terrestrial climates or paleocirculation patterns of the atmosphere and surface waters of the oceans. The extensive geographic coverage of core samples and the refinement of quantitative approaches to paleoclimatology (e.g., CLIMAP, 1976) have demonstrated the value of such techniques in interpretation, and perhaps in prediction as well.

By contrast, our understanding of the abyssal circulation of the world's oceans and its variations during the past is relatively poor. There are at least two major factors which have contributed to our difficulty in attempts at interpretation:

(1) One must first identify appropriate "base-level" conditions. In other words, one must establish unequivocally a one-to-one correspondence between parameters in the surface sediments and a characteristic property of the overlying water.

(2) One must assume that the physical properties and flow characteristics of the near-bottom water have been sufficiently stable over time such that they remain identifiable in the geologic record when integrated over thousands of years.

It is of course highly unlikely that the abyssal circulation has remained perfectly uniform *anywhere* on a geologic time scale. Nevertheless, meaningful paleo-oceanographic interpretations of the abyssal circulation require that this condition be met as closely as possible.

The Vema Channel in the southwestern Atlantic is one of several topographic gaps which plays a major role in controlling the abyssal circulation of the world's oceans. Other similarly important features include the Samoan Passage (Hollister et al., 1974; Reid and Lonsdale, 1974; Johnson, 1974a), Kane Gap (Hobart et al., 1975), Walvis Gap (Connary, 1972), and numerous fracture zones in the mid-oceanic ridge system (e.g., Gibbs Fracture Zone, Romanche Trench). These passages are of particular significance for paleocirculation studies in two principal respects:

(1) Bottom water flow through these passages may be relatively uniform in comparison with that observed in more open regions of the abyssal ocean. Directional deviations in the flow are generally minimal due to topographic constraints, and the high-frequency tidal components to the flow are sometimes (though not always) overshadowed by a strong unidirectional component representing net transport between larger basins via these passages (e.g., Reid and Lonsdale, 1974). If in fact the bottom boundary layer in these regions is relatively time-independent, then sediment—current interactions may also be relatively stable over time. Consequently, there is relatively high assurance that sea floor properties as observed in bottom photographs and sediment samples *are* in fact in equilibrium with the existing flow conditions.

(2) Interpretations of paleocirculation must make the additional assumption that flow characteristics were sufficiently stable such that they remained identifiable in the geologic record when integrated over thousands of years. Passages such as the Vema Channel may provide the most suitable type of region for satisfying this condition, assuming that the first-order basin/basin transport of bottom water is uniform over geologically significant time periods.

REGIONAL SETTING OF VEMA CHANNEL

Previous investigations have shown that Antarctic Bottom Water (AABW) dominates the abyssal circulation of the southwestern Atlantic (Wüst, 1957; Gordon, 1972; Reid et al., 1973). Upon formation in the Weddell Sea, a major portion of this water mass flows northward around the Scotia Arc and through a topographic gap in the Falkland Fracture Zone, near 49° S and 36° W (Le Pichon et al., 1971a). Upon leaving this gap, AABW flows to the west along the northern edge of the Falkland Plateau. As it approaches the continental rise in the extreme southwestern portion of the Argentine Basin, the bottom current is deflected toward the north and continues as a deep western boundary current (Fig.1).

At the northern margin of the Argentine Basin the flow of AABW is restricted by the Rio Grande Rise, which forms a topographic barrier

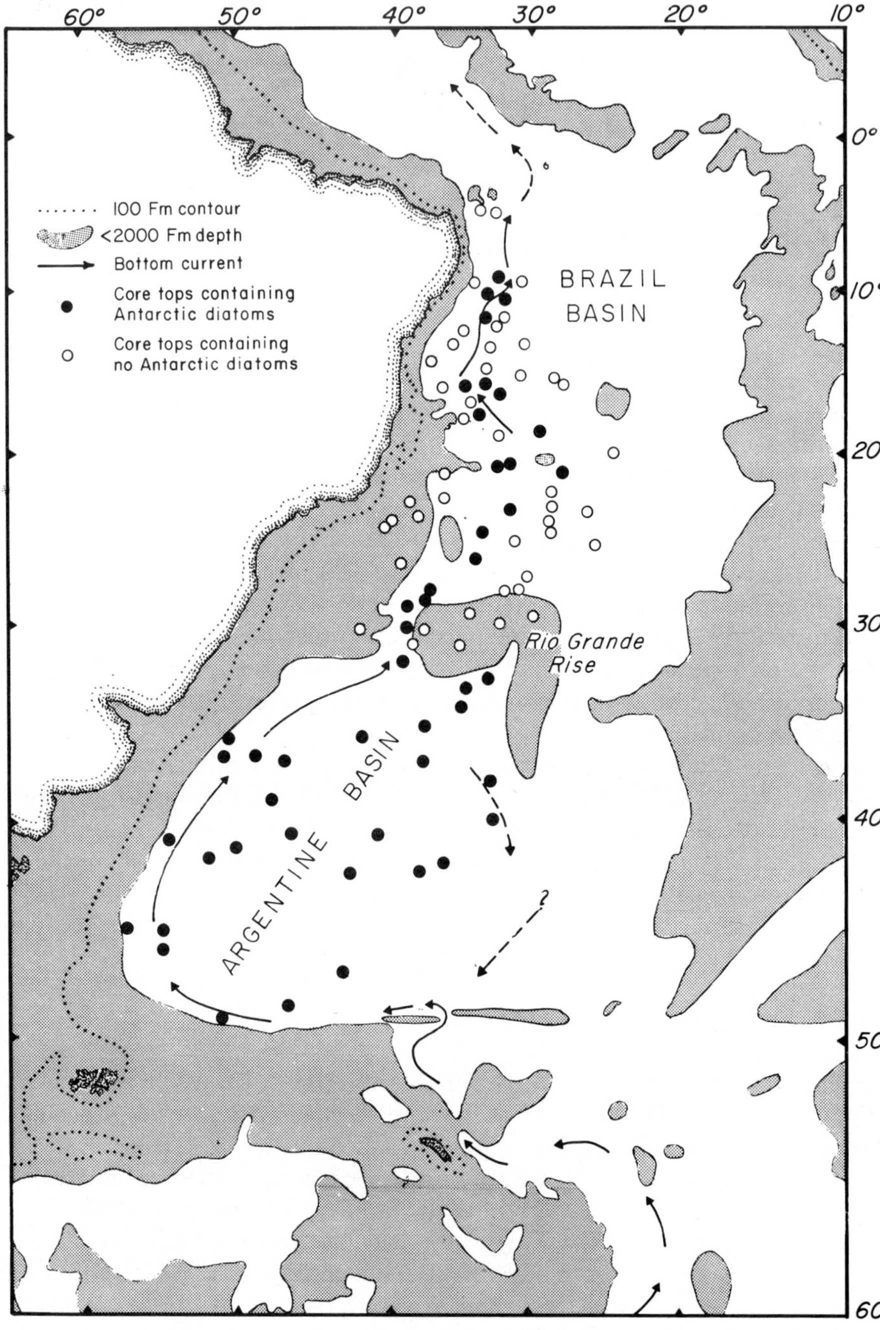

Fig. 1. Regional bathymetry of the southwestern Atlantic. Shaded region designates areas shallower than 2000 fathoms. Arrows designate the direction of flow of Antarctic Bottom Water. The Vema Channel, whose bathymetry is shown in Fig. 3, separates the Rio Grande Rise from the continental margin of South America.

between the Brazil and Argentine Basins (Fig.1). The principal deep passage (greater than 4500 m) through the Rise is the narrow, sinuous gap referred to as the Vema Channel (Le Pichon et al., 1971b). Numerous investigations (Wright, 1970; Le Pichon et al., 1971b; Reid et al., 1973; Johnson et al., 1976) have documented the presence of a significant northward transport of AABW through the Vema Channel and into the Brazil Basin, with maximum current velocities on the order of 20—25 cm/sec (Johnson et al., 1976).

The Vema Channel is centered near 39° 30′W, with a sill depth of approximately 4600 m (Lonardi and Ewing, 1971; this report, Figs.2 and 3). Seismic reflection profiling (LePichon et al., 1971b) has demonstrated that the basement relief in the vicinity of the channel is highly irregular, and that these irregularities may have been influential in controlling the orientation and morphology of the channel during its development. In cross-section the main branch of the channel is notably asymmetrical (Fig.4, profile *AB*). The deepest part of the channel is generally adjacent to the western wall, a high and steep slope which on many crossings appears to represent outcrops or near-outcrops of acoustic basement (Le Pichon et al., 1971b; Figs.2, 3). On the eastern side of the channel the walls rise steeply to a depth of ~4200 m and then level out onto a broad terrace up to ~100 km in width (Fig.3). The terrace is underlain by >1 km of acoustically transparent sediment of unknown lithology and age. The eastern margin of the terrace merges imperceptibly with the lower flanks of the Rio Grande Rise.

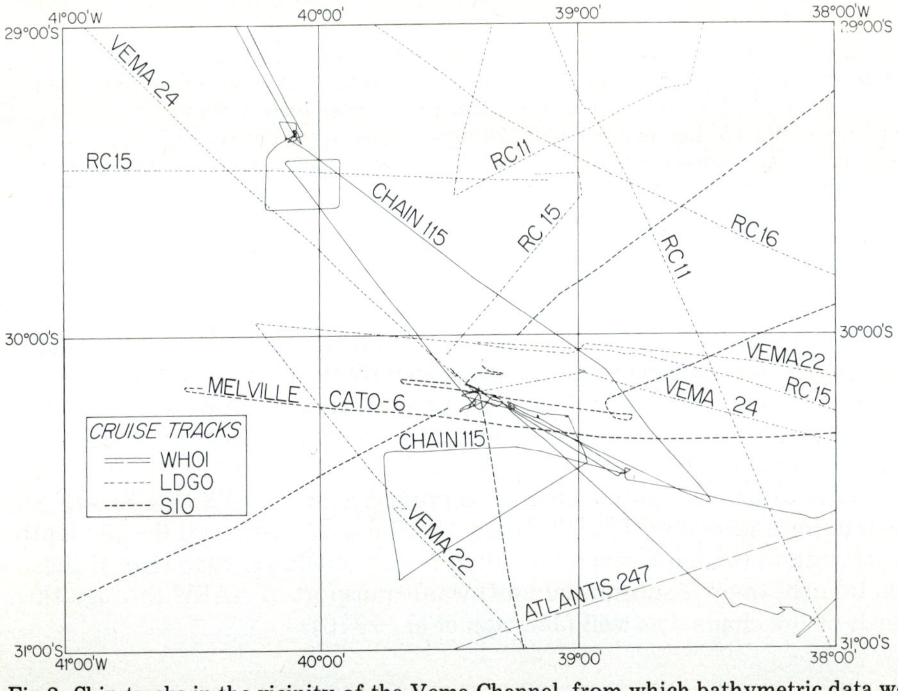

Fig.2. Ship tracks in the vicinity of the Vema Channel, from which bathymetric data were used in constructing a revised bathymetric chart (Fig.3).

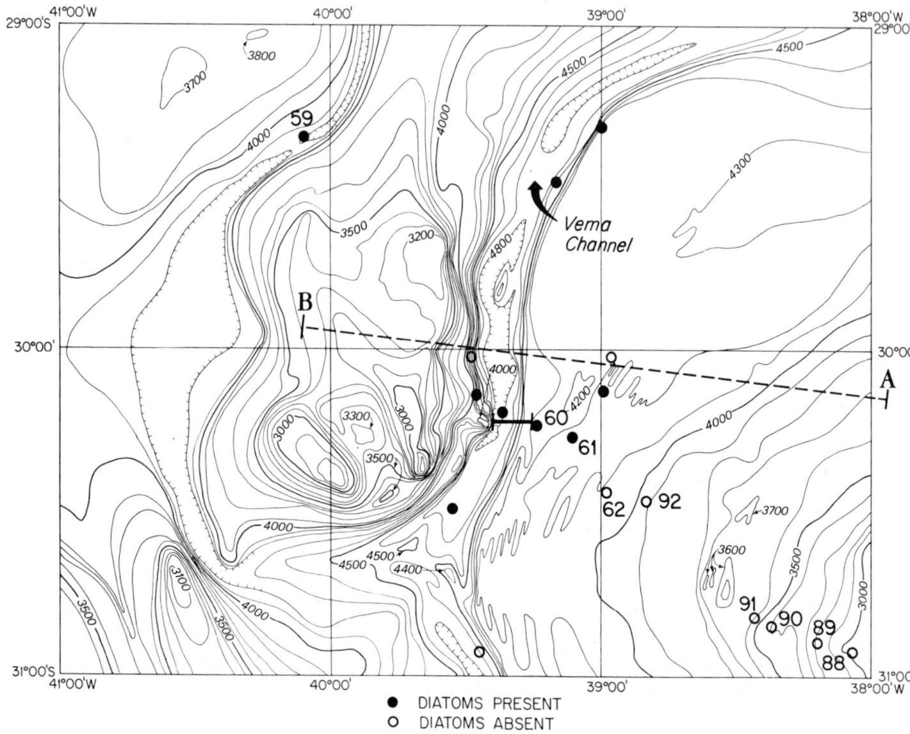

Fig.3. Bathymetry of the main branch and western branch of the Vema Channel, based on data from tracks shown in Fig.2. Elsewhere in the region the contours of Lonardi and Ewing (1971) have been retained. Contours are in corrected meters. Numbered circles show the locations of the nine "Chain" 115 piston cores in the channel vicinity; unnumbered circles represent LDGO cores which were examined. Filled circles designate cores in which core-top samples contained Antarctic diatoms. Dashed line labeled *AB* indicates the location of seismic profile *AB* shown in Fig.4. Heavy line crossing the channel axis at 30° 13′S shows the location of the temperature and nephelometer profiles of Fig.5.

The asymmetrical topography and structure of the channel suggests that on a geological time scale there has been preferential erosion and perhaps strongest current velocities adjacent to the western wall. However, there are insufficient direct current observations to demonstrate this interpretation (Johnson et al., 1976).

A major bifurcation in the channel is present near 30° 40′S (Fig.3), at which point a secondary branch diverges to the west. Although the sill depth in this western branch is several hundred meters shallower than that of the main branch, there is substantial northward transport of AABW through this branch of the channel as well (Johnson et al., 1976).

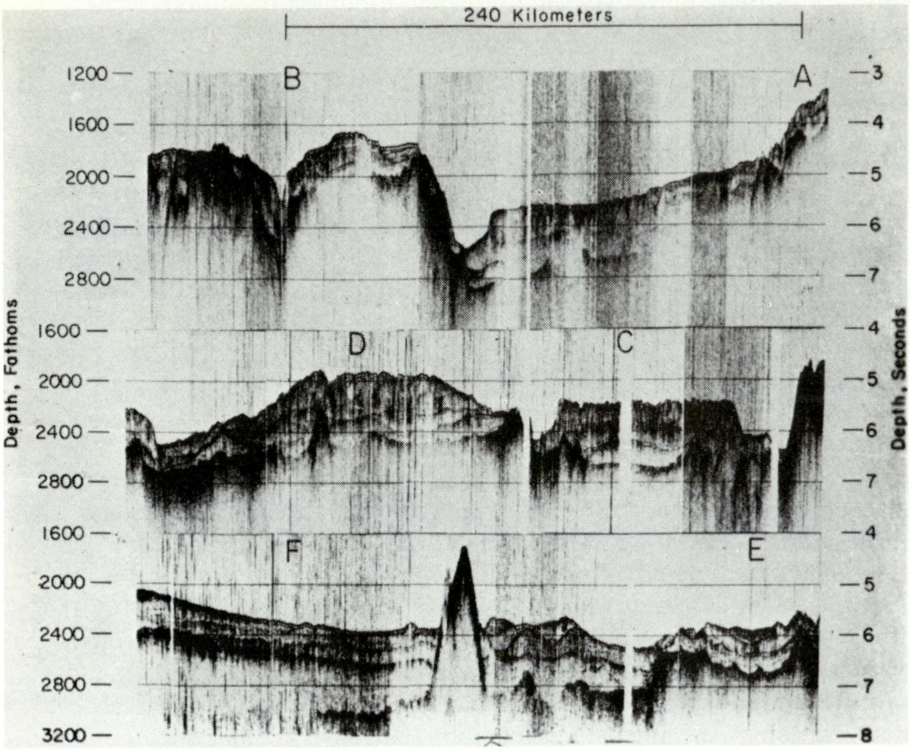

Fig.4. Seismic reflection profiles across the axis of the Vema Channel, modified after Le Pichon et al. (1971b). Profile *AB* crosses the channel axis at the location shown on Fig.3. Profiles *CD* and *EF* cross the channel to the south of the study area. Note the topographic and structural asymmetry of the channel region, indicating that preferential erosion has occurred adjacent to the western wall. The broad terrace on the eastern flank corresponds approximately with the upper limit of AABW.

OBJECTIVES AND METHODS OF STUDY

During Leg 6 of Cruise "Chain" 115, geological, geophysical and physical oceanographic observations were undertaken in the Vema Channel region in view of the following principal objectives:

(1) Define the present-day abyssal hydrography, circulation, and benthic boundary layer structure within the channel and on the lower flanks of the Rio Grande Rise.

(2) Identify well-defined parameters in the surface sediment which reflect the present-day hydrographic regime.

(3) Examine the Pleistocene record in sediment cores from the east flank of the Vema Channel in order to identify, and establish a chronology for, possible changes in abyssal circulation through the channel during the past million years.

In view of these objectives the following shipboard operations were carried out.

Bathymetry

The track of R/V "Chain" on cruise 115 is shown in Fig.2. More detailed navigation plots are presented in the cruise report (Johnson, 1974b). Echo-sounding (3.5 kHz) and continuous seismic reflection (air gun) profiles were obtained during underway operations, and were used in conjunction with other data (Fig.2) to construct a revised bathymetric chart of the Vema Channel region (Fig.3).

Hydrography

Abyssal temperature profiles were obtained at ten stations in the Vema Channel region (Johnson et al., 1975), using conventional hydrocasts together with thermograd profiles from a core-mounted heat flow recorder. Profiles of light scattering were taken at eight locations, and an ocean-bottom nephelometer was deployed in the channel axis to record temporal variations in the suspended load of AABW (Johnson and Sullivan, 1976). Five free-vehicle current meters (Schick et al., 1968) were deployed to measure deep and bottom water flow in the vicinity of the Vema Channel over a 15- to 17-day period. These data have been synthesized into an interpretation of the abyssal hydrography, currents, and benthic boundary layer structure in the Vema Channel region (Johnson et al., 1976).

Sediments

Nine piston cores were obtained during "Chain" 115 in the Vema Channel region (Fig.3; Table I). One (core 59) was located in the axis of the western branch, and eight were along a profile extending from 2941 m to 4310 m on the lower flank of the Rio Grande Rise. Additional core top samples from LDGO cores in the region (Table I) were examined in order to characterize the surface sediment lithology in terms of parameters which may be diagnostic of AABW. The eight cores in the "Chain" 115 profile (Fig.3) were positioned so as to complement those cores obtained along the same profile by "Charcot" (Melguen and Thiede, 1974). Subsequent work on the "Chain" 115 cores in conjunction with the "Charcot" samples should lead to further refinements of the interpretations presented previously (Melguen and Thiede, 1974; Diester-Haass, 1976) and in this report.

Laboratory analyses of the core material were undertaken as follows. All cores were split, photographed, described visually, and analyzed microscopically (smear slides) at 1-m intervals and at all major lithologic boundaries. Total carbonate was measured down-core at 10-cm intervals. Principal sediment types were designated on the basis of these analyses. Additional studies were carried out according to the following procedures:

TABLE I

WHOI and LDGO piston cores from the Vema Channel region which were examined during this study

Core number	Latitude (S)	Longitude (W)	Water depth (m)	Core length (cm)
CH 115-59	29° 20.8′	40° 05.8′	4188	753
CH 115-60	30° 13.8′	39° 14.6′	4310	547
CH 115-61	30° 15.5′	39° 05.8′	4181	819
CH 115-62	30° 24.6′	38° 58.4′	4065	712
CH 115-88	30° 55.0′	38° 04.8′	2941	704
CH 115-89	30° 52.8′	38° 11.8′	3151	672
CH 115-90	30° 51.0′	38° 22.3′	3384	654
CH 115-91	30° 49.5′	38° 25.8′	3576	558
CH 115-92	30° 25.7′	38° 50.3′	3934	734
RC11-40	29° 18.5′	39° 00.0′	4299	1091
RC15-147	29° 29.8′	39° 09.7′	4664	1145
RC15-148	30° 01.8′	39° 27.3′	4201	576
V16-188	30° 56.0′	39° 27.0′	4665	10
V22-74	30° 02.0′	38° 57.0′	4235	966
V22-75	30° 31.0′	39° 33.0′	4803	352
V24-249	30° 07.0′	38° 59.0′	4184	1046
V24-250	30° 11.0′	39° 22.0′	4813	382
V24-251	30° 08.0′	39° 28.0′	4111	382

Particle size analysis. Samples of approximately 6 cc were washed in distilled water and disaggregated in a Calgon solution before being wet-sieved through a 62-μm sieve. The fraction finer than 62 μm (4 ϕ) was allowed to settle before decantation. After a second settling, the fraction containing particles in the size range 4—8 ϕ was analyzed in a Model ZB Coulter Counter utilizing a 200-μm aperture. The Coulter Counter method of size analysis is based on particle volume, and has been utilized in measurements of fine-grained materials in a wide variety of applications (McCave and Jarvis, 1973; Walker et al., 1974; Huang et al., 1975). A more detailed explanation of the technique is presented by Swift et al. (1972). The method employed in our analyses allowed interpretation of the particle size distribution with particle size increments of 0.33 ϕ.

Shear strength measurements. The undrained shear strength was measured with a cone penetrometer at 10-cm intervals on the archive half of the split core. The cone penetrometer is a stainless steel cone which is allowed to free fall from a constant height into the sediment. The depth of penetration is proportional to the shear strength of the soil and can be determined from calibration curves for the cone in use. Shear strength and cohesion are equivalent in fine-grained cohesive soils if the test is applied without loss of pore water.

Foraminiferal dissolution indices. Samples were analyzed at 10-cm intervals in three cores (88, 90, 92) in which well-defined Late Pleistocene carbonate cycles were identified. The coarse fraction (>250 μm) was separated, and counts of 300 to 500 individual forams per sample were made. Each individual was counted as either a fragment or a whole test, and the ratio P_f/P_w was interpreted as an index of the relative dissolution down-core. The effects of selective dissolution upon individual species are being investigated separately (G. Lohmann and L. Tjalsma, in preparation) and will not be considered here.

Antarctic diatoms. Samples were taken at 25-cm intervals from seven of the cores (59, 60, 61, 62, 88, 89, 92), and were prepared for analysis of the presence of diatoms following the methods of Schrader (1974). One half gram of oven-dried sediment was placed in a 1:1 solution of acetic acid and hydrogen peroxide, and warmed for a 20-minute period. The coarse fraction was removed by allowing the sample to remain in suspension in a 10 cm high column of water and then decanting. This procedure was repeated several times. The sample was washed with repeated centrifuging. Finally the washed residue was suspended in 50 ml of de-ionized water and 0.3-ml aliquot was removed and placed on a 18 × 18 mm cover slip. After drying, this cover slip was bonded to a glass slide using Carmount as the mounting medium (*r.i.* = 1.65).

Paleomagnetic stratigraphy. Paleomagnetic sampling was carried out with 6-cc plastic boxes pushed into the core at 10-cm intervals and withdrawn after careful orientation. Selected samples were demagnetized in alternating magnetic field in steps of 50 oersteds in order to determine the peak demagnetizing field. All samples were then demagnetized in the peak field of 100 oersted prior to measurement of the remnant magnetism. Direction and intensity of the remnant magnetism were measured on a Digico Balanced Fluxgate Magnetometer. The vertical component of remnant magnetism was used to determine the paleomagnetic polarity (see Fig.10). In the southern hemisphere the inclination is negative (down) when normal and positive (up) when reversed.

Nannofossil biostratigraphy. Smear slides were examined to determine an approximate biostratigraphic age for all carbonate-bearing intervals of the cores. All cores in which nannofossil assemblages are present (i.e., all except core 60) failed to reach pre-Pleistocene sediments. Reworked Tertiary discoasters were rare to common in the four deepest cores (59—62); the presence of reworked assemblages in these cores, however, precludes the reliable use of nannofossil biostratigraphy to subdivide the Pleistocene. In the five shallowest cores (88—92) no evidence of reworking was encountered at any levels. Stratigraphic control within these cores has been proposed using oxygen isotope techniques and core-to-core correlation of the Brunhes carbonate cycles.

Oxygen isotope stratigraphy. Bulk foraminiferal assemblages (>62 μm) were analyzed at 10-cm intervals in the two shallowest cores (88 and 89), in an attempt to establish a reliable chronology for the Late Pleistocene carbonate cycles in these and other cores. The foram assemblages at all levels in both cores analyzed were relatively uniform in species composition, and were dominated by two taxa which are relatively susceptible to dissolution, *Globigerinoides ruber* and *G. sacculifer*. We therefore are reasonably confident that the down-core variations in $\delta\ ^{18}O$ reflect principally global ice volume changes and not selective dissolution of the assemblages. A complete tabulation and discussion of all isotopic analyses is presented in a separate data report (Peters, 1976). Confirmation of the interpretations presented here will require analysis of monospecific assemblages, inasmuch as isotopic fractionation among foraminifera is apparently species-dependent.

ABYSSAL HYDROGRAPHY AND AABW FLOW

Recent investigations have led to a more precise interpretation of the abyssal hydrography and flow regime within the Vema Channel (Reid et al., 1973; Johnson et al., 1976). These results and their geological implications may be summarized as follows:

(1) Profiles of abyssal temperature and light scattering within the channel axis (near 30°13′S) show sharp gradients of both parameters in the transition zone between AABW and the overlying NADW (Fig.5). This transition zone, however, is notably asymmetrical across the channel axis. On the eastern side of the channel, the benthic thermocline is sharp (0.8°C/100 m to 1.7°C/100 m) and relatively deep (~4250 m). In the channel axis adjacent to the western wall, the benthic thermocline is several hundred meters shallower and more gentle (~0.3°C/100 m). Gradients in light scattering exhibit a corresponding asymmetry. Coldest bottom water temperatures ($\theta < -0.18°C$) and highest values of near-bottom turbidity are present on the eastern side of the channel. Frictional interaction at the bottom boundary layer apparently leads to eastward (upslope) Ekman-layer advection of the coldest water away from the channel axis.

(2) Hydrographic data suggest that the upper boundary of Antarctic Bottom Water in the channel region may correspond approximately with the density surface $\sigma_4 = 46.00$, or the potential temperature isotherm of 1.2°C (Johnson et al., 1976). This interface lies below lower NADW ($\sigma_4 = 45.93$) and near the top of the benthic thermocline and bottom nepheloid layer (Fig.5). The interface slopes downward toward the east from ~3770 m in the channel axis to ~4000 m on the lower flanks of the Rio Grande Rise. Eastward extrapolation of hydrographic data from the channel axis to the flanks of the rise should take into account this effect of downward-sloping isopycnals and isotherms.

(3) Direct current measurements and geostrophic computations indicate the presence of relatively strong currents (20—25 cm/sec) in the region of AABW flow (Johnson et al., 1976). Since these current speeds may be

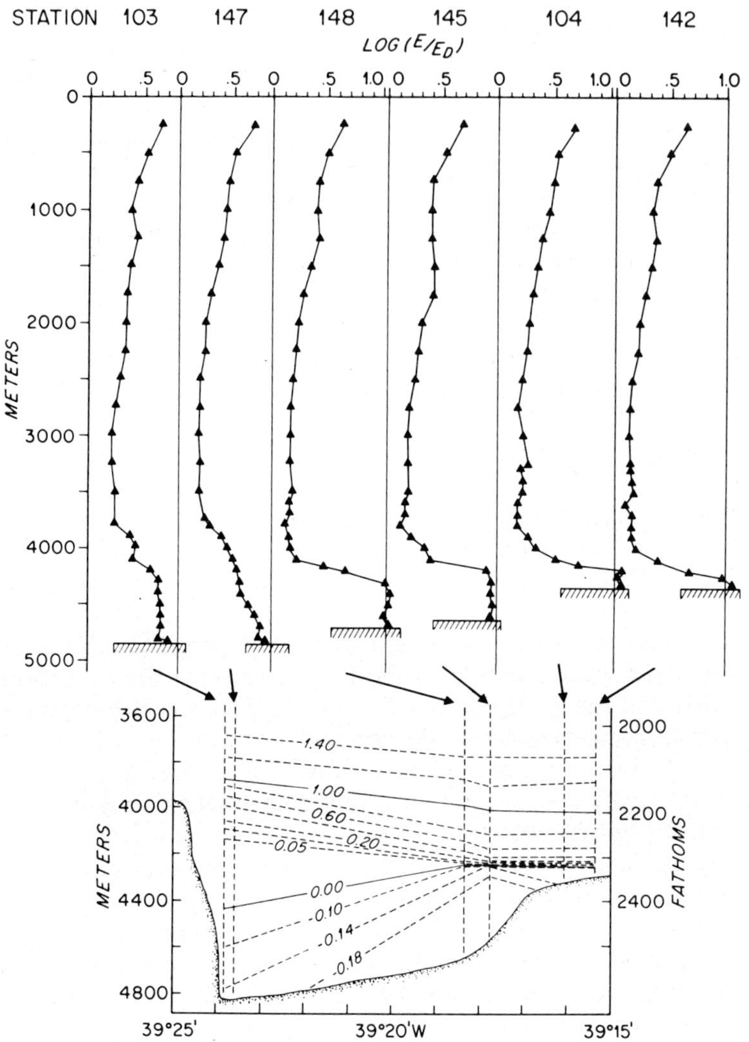

Fig.5. Light scattering profiles (upper) and abyssal temperature observations (lower) in a cross-section of the Vema Channel axis near 30° 13′S. Location of section is shown on Fig.3. Note the cross-sectional asymmetry in both the benthic thermocline and bottom nepheloid layer. The upper boundary of Antarctic Bottom Water corresponds approximately with the 1.2° C potential temperature isotherm (Johnson et al., 1976).

sufficient to erode deep-sea hemi-pelagic clay, particularly in the presence of microrelief such as Mn nodules (Lonsdale and Southard, 1974), depositional unconformities would be likely within the region of the sea floor underlying AABW. Consequently care is required in stratigraphic interpretations of cores from these deeper regions and in the use of techniques (such as paleomagnetic stratigraphy and isotopic analysis) which rely upon the assumption of uninterrupted deposition.

(4) From a geological standpoint it is important to determine the time-variability of the depth and gradient of the benthic thermocline, and whether there is a corresponding fluctuation in the position of the NADW/AABW transition zone on the east flank of the channel. Comparison of hydrographic data from 1972 ("Geosecs" 59) and 1974 ("Chain" 115) in the channel axis suggests that the benthic thermocline may migrate vertically by up to 200—300 m on a time scale of months to years. A corresponding variability in the upper boundary of AABW would be likely on the lower flanks of the Rio Grande Rise. However, the deposition rates in the Vema Channel cores ($\sim$1 cm/10^3 years) and the sample size used in this study ($\sim$2 cm length) are such that episodic events of relatively short duration (i.e., less than $\sim 10^3$ years) would probably not be resolvable in the geologic record.

SURFACE SEDIMENTS

The tops of 18 cores from the Vema Channel region (Table I; Fig.3) were examined in an attempt to identify parameters which may be diagnostic of various hydrographic conditions and/or depositional environments. These cores may be subdivided geographically as follows:

Floor of the Vema Channel

Four of the cores are from depths greater than 4500 m near the channel axis. All core-top material at these locations consists of carbonate-free clay, with common manganese nodules, encrustations, and micronodules. The coarse fraction consists dominantly of detrital mineral grains, micronodules, and very rare siliceous skeletal debris. No calcareous material is present.

Western branch of channel

Core 59, from the axis of the western branch (Fig.3), consists of nannofossil clay, with common manganese nodules and micronodules. Foraminifera are present in the coarse fraction. The cold bottom water temperature ($\theta < 0°$C) in the western branch documents the presence of AABW here (Johnson et al., 1976). Yet the presence of a significant carbonate component in the sediment suggests that the corrosive effects of the water upon the carbonate fraction are substantially less than in the main branch.

Main channel, western wall

Two cores from the steep western wall (RC15-148; V24-251) contain strikingly different surface lithologies, a contrast which may reflect differential solution above and below the NADW/AABW transition zone. Core 251, from $\sim$4100 m (Table I), consists of foraminiferal ooze. Core 148, from 4200 m, contains carbonate-free clay, with only micronodules in the coarse fraction.

Main channel, east flank

Seven of the cores examined extend through a depth range of 4000—4500 m on the east flank of the Vema Channel (Fig.3). The surface sediment in each of these cores is a calcareous clay to calcareous ooze, which grades downward into an unfossiliferous clay. Foraminifera are abundant in the shallower cores from this group. The coarse fraction, total carbonate, and manganese micronodules all decrease in abundance downward in these cores.

Lower flank of Rio Grande Rise

Five cores (88 through 92) extend up the lower flank of the Rio Grande Rise above a depth of 4000 m. The surface sediment in each of these cores is a calcareous ooze, with abundant foraminifera and nannofossils.

Core-top samples from the eight "Chain" 115 cores on the east flank were examined to determine the depth intervals within which the carbonate fraction changes composition. The following three levels were identified (Fig.6):

Pteropod compensation depth (~3200 m). Pteropods are common in the coarse fraction of core 88, very rare in core 89, and absent below core 89.

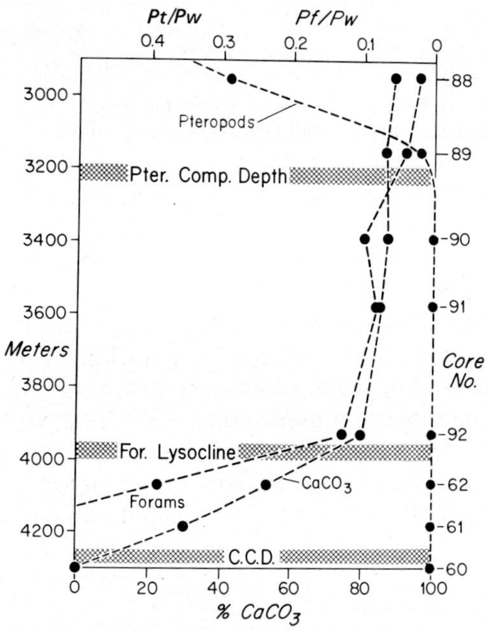

Fig.6. Dissolution characteristics of the calcareous component of core top samples from the east flank of the Vema Channel. The foraminiferal lysocline corresponds to θ = 1.2° C, and may be an appropriate criterion for identifying the upper limit of AABW in the channel region (Johnson et al., 1976). The CCD corresponds with the maximum gradient in the benthic thermocline, and with a depth of rapid deposition of fine-grained sediment.

We can therefore designate a depth slightly below core 89 as an appropriate pteropod compensation depth. This estimate compares favorably with that of Melguen and Thiede (1974).

Foraminiferal lysocline (~4000 m). A foraminiferal dissolution index was computed for each core-top sample, based on counts of 300 to 500 specimens per sample. The dissolution index, P_f/P_w, is the ratio of foram fragments to whole tests in each sample. For cores 88 through 92, the dissolution index is approximately 0.10 or less. In the next deepest core (62), the dissolution index is ~0.40, and below core 62 foraminifera are absent. Consequently it appears that a depth of 4000 m (between cores 92 and 62) is an appropriate choice for the position of the foraminiferal lysocline on the east flank of the channel. This depth corresponds to the 1.2°C potential temperature isotherm and the density surface $\sigma_4 = 46.00$, which may be an appropriate choice for defining the transition zone between NADW and AABW (Johnson et al., 1976).

Carbonate compensation depth (~4250 m). Total carbonate in core-top samples decreases from 80% in core 92 to 53% in core 62, 30% in core 61, and 0 in core 60. Therefore the compensation depth for $CaCO_3$ lies between cores 60 and 61 or at approximately 4250 m. This depth corresponds approximately with the maximum gradient in near-bottom temperature and light scattering on the east flank of the channel (Fig.5; Johnson et al., 1976).

Antarctic diatoms in core top samples can be used to trace the path of spreading of the AABW. Fig.1 is a summary of data from more than 60 core tops in the southwest Atlantic (Burckle, 1976). The distributions clearly show that the AABW has been the principle agent in moving Antarctic diatoms from their high southern latitude source. Fig.3 is an enlargement of the Vema Channel area. Diatom-bearing sediments are confined to the lower slope of the Vema Channel below approximately 4100 m. This is approximately correlative with the top of facies IV of Melguen and Thiede (1974). These authors report diatom fragments only from their facies IV and V, both of which are within the AABW. The absence of diatoms in samples from the axis of the Vema Channel may be attributed to intense circulation preventing these small particles from being deposited. Indeed, many samples from the floor of the channel contain a manganese crust with no biogenic component present.

LITHOLOGY AND STRATIGRAPHY OF VEMA CHANNEL CORES

Major lithologic types

Fig.7 summarizes the lithology of the nine "Chain" 115 piston cores from the Vema Channel region. The cores can be conveniently subdivided, using lithological criteria, according to their geographic locations relative to the AABW/NADW transition zone. Cores 88 through 92 occur within the range of present-day NADW, whereas the remaining cores lie within the depth interval presently occupied by AABW.

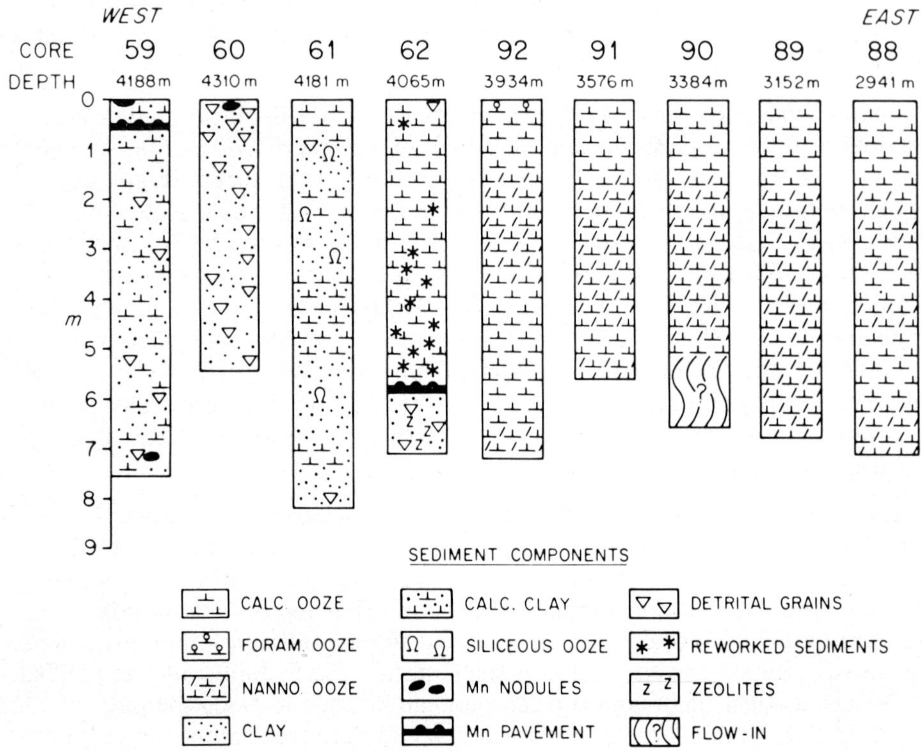

Fig.7. Lithology of "Chain" 115 cores from the Vema Channel area. Core locations (numbered) are shown in Fig.3.

Cores 88 through 92 consist of alternating units of calcareous ooze and nannofossil ooze, with the total carbonate content varying between 55% and 90% (Fig.8). Relatively coarse units rich in planktonic foraminifera alternate with finer-grained units in which nannofossils predominate. All microfossil assemblages are of Pleistocene age, and there is no admixture of reworked older forms. Detrital mineral grains and manganese micronodules constitute a negligible portion of the coarse fraction, and are present only in small percentages among the smear slide components. The near-uniform visual appearance of the cores precludes reliable subdivision without microscopic examination and chemical analysis. In all cores the carbonate cycles are well defined, and appear to have a periodicity on the order of 100,000 years, based on the position of the Brunhes/Matuyama boundary (Fig.8). There is no evidence for significant erosion or non-deposition in these cores.

In the four cores from below 4000 m the lithologic units are more varied, and the stratigraphic continuity within the cores is more open to question. Cores 59, 60 and 61 are relatively barren of carbonate, and consist principally of hemipelagic clay, detrital mineral grains, and manganese micro-

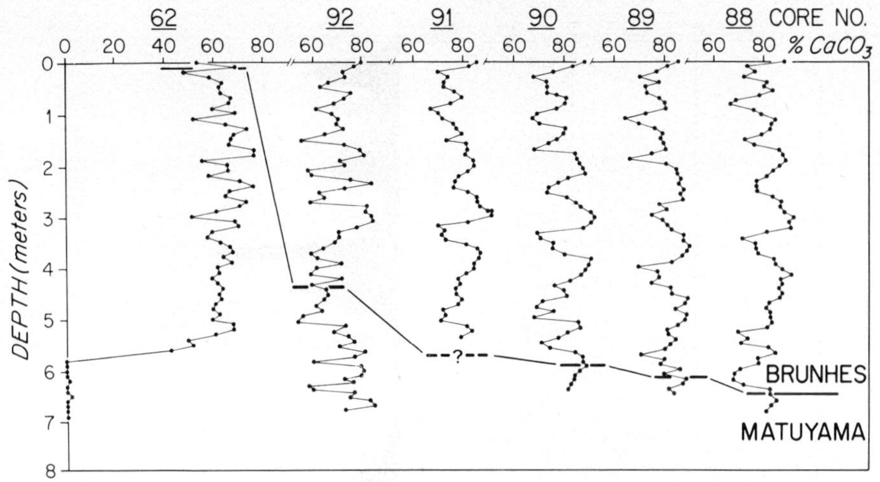

Fig.8. Percent calcium carbonate in six piston cores from the east flank of the Vema Channel. Position of the Brunhes/Matuyama polarity boundary is shown.

nodules (Fig.9). Core 62 contains calcareous ooze down to ~6 m at which level a manganese pavement is present. Below this pavement is an unfossiliferous clay. Core 60 is totally devoid of carbonate, whereas cores 59 and 61 contain alternating units of high- and low-carbonate material, with carbonate-rich sediment at the sea floor and carbonate-poor material below. Varying proportions of reworked Tertiary discoasters are present in some intervals of core 62.

Magnetic stratigraphy

The paleomagnetic polarity was determined for each of the nine "Chain" 115 cores in the Vema Channel (Fig.10). The polarity was interpreted from the direction of inclination of remnant magnetization. Fig.11 presents the polarity interpretation for one of the cores (No. 92), and the magnetic intensity, declination, and inclination data upon which the polarity interpretation is based. Similar data were obtained for the remaining eight cores, but only the polarity interpretations are included in this report (Fig.10).

In the five shallowest cores on the east flank of the channel (Nos. 88—92), a thick section of sediment of normal polarity is present at the top of each core. This interval is interpreted to be of Brunhes age, and decreases systematically in thickness with increasing water depth between core 88 and core 92 (Fig.10). The decrease in thickness of Brunhes sediment corresponds with a decrease in the mean carbonate content of the same cores (Fig.8). The close correspondence between the carbonate cycles in these five cores supports this interpretation of the paleomagnetic stratigraphy.

In the four deepest cores (Nos. 59—62) the magnetic polarity records are irregular, and stratigraphic interpretation is less certain than in the five

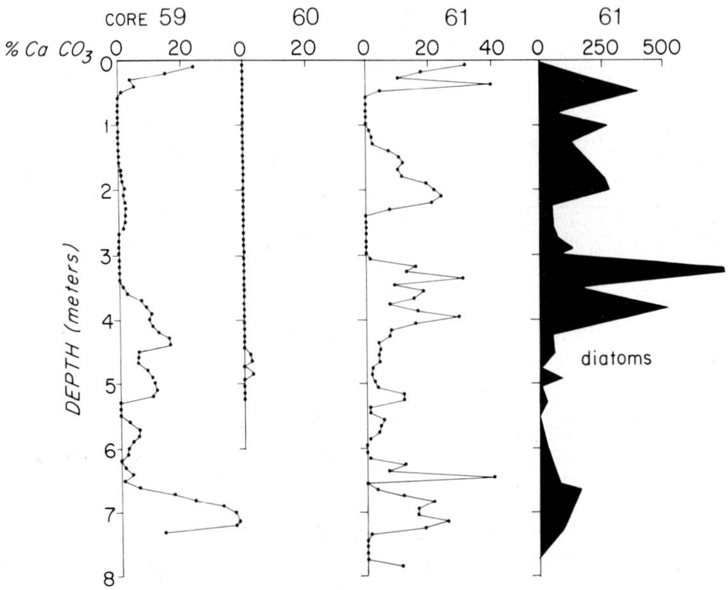

Fig.9. Percent calcium carbonate in three piston cores from the western branch (core 59) and main axis (cores 60 and 61) of the Vema Channel. The variability in carbonate content in cores 59 and 61 may be a consequence of variable rates of dissolution and episodic bottom water flow. The relative abundance of Antarctic diatoms in core 61 is indicated, in units of thousands of specimens per gram of sample.

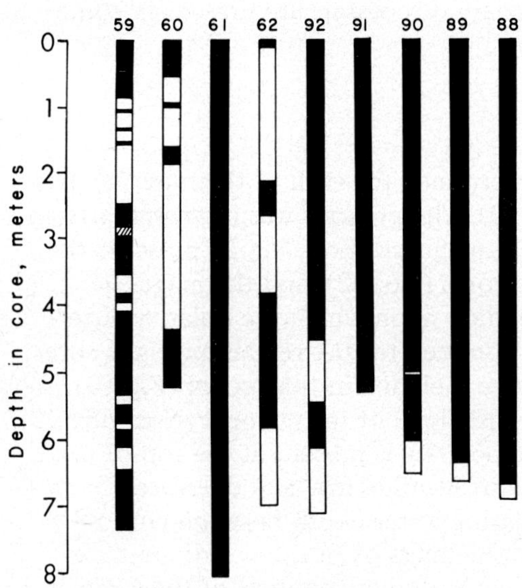

Fig.10. Paleomagnetic polarity logs of "Chain" 115 cores from Vema Channel. Black indicates normal polarity, white is reversed polarity, and cross-hatched is of transitional polarity. Core locations are shown in Fig.3.

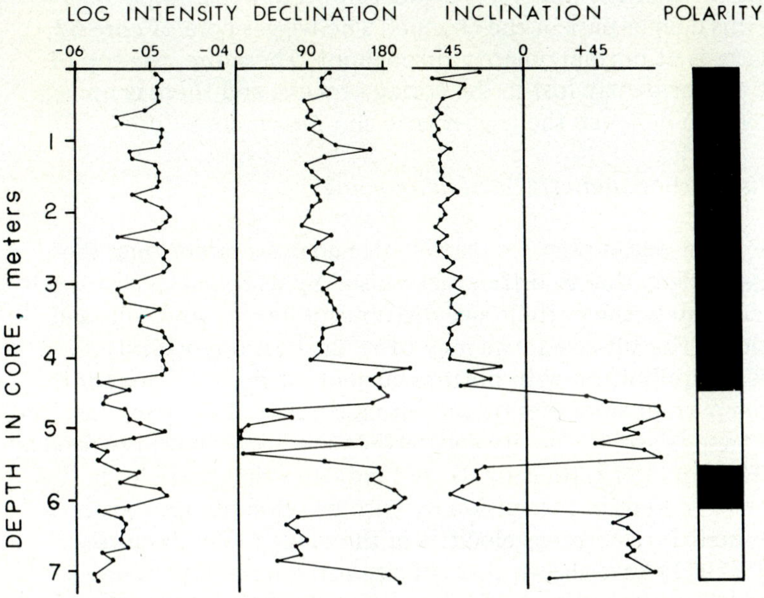

Fig.11. Plot of paleomagnetic intensity, declination, and inclination after 100 oersted peak alternating field demagnetization of core CH 115-6-92. Paleomagnetic polarity log was inferred from inclination data only. In the polarity log black indicates normal and white reversed polarity.

shallower cores. Core 62 is of Pleistocene age down to the manganese pavement at 6 m; consequently the most likely interpretation of the magnetic polarity record is that the normal interval above the manganese pavement (4—6 m) corresponds to the Olduvai event near the Pliocene/Pleistocene boundary. Core 61 is of normal polarity throughout, and therefore apparently failed to penetrate the Brunhes/Matuyama boundary. In core 60 there is only a thin (<1 m) interval of normal polarity sediment at the top of the core, suggesting that Brunhes-age sediment is thin or perhaps missing entirely from this site. The absence of microfossils has so far precluded the identification of the magnetic polarity boundaries in core 60. The polarity record of core 59 (Fig.10) is difficult to interpret, and may be a consequence of effects of bioturbation at this location. The deepest carbonate-bearing sediment in core 59 is of Pleistocene age, and there appears to be no reliable way to interpret the polarity log of core 59 with the Pleistocene paleomagnetic time scale.

An attempt was made to verify the reality of the anomalously short section of normal polarity sediment at the top of cores 60 and 62 (Fig.10) by comparison with the polarity of the trigger core. The polarity stratigraphy of piston core 60 is consistent with that of the trigger core, which suggests that the short normal polarity section at the top of piston core 60 is real. Even if this short section of normal polarity were demonstrated to be

Brunhes, the thinness of the interval is evidence of scour or intense winnowing in this deeper part of the channel. The trigger core of core 62, 140 cm in length, is of normal polarity throughout. Therefore, the top of piston core 62 was evidently lost in the coring process, and there is no stratigraphic overlap between the trigger core and piston core.

Particle size distribution and effects of winnowing

The silt size range was chosen for particle size analysis rather than the complete range of sizes, due to difficulties caused by using various methods to determine separately the particle size distribution in the sand, silt, and clay size fractions. The silt size range may offer the best opportunity to study particles in equilibrium with bottom currents at the sea floor for the range of bottom current velocities present in most areas of the deep sea. The relationship between current velocity and particle size has been summarized by Hollister and Heezen (1972, their fig.3), and indicates that particles in the silt size range will be kept in suspension by current velocities greater than 1 cm/sec, and eroded by current velocities in the range of 6—15 cm/sec. Southard et al. (1971) have shown that silt size foraminiferal tests are eroded by currents in the range of 15—35 cm/sec. Therefore, particles of silt size may act as sensitive monitors of changes in bottom water velocity by means of sediment textural parameters.

As currents winnow or scour fine sediment, the particle size distribution is changed and recorded in textural parameters such as mean grain size, skewness, and sorting. An increase in current velocity will result in an increase in mean grain size, a more positive skewness, and better sorting (Huang and Watkins, 1977). By analyzing zones within a core where these conditions are met, it may be possible to identify horizons of increased winnowing or zones of scour which could go undetected in the presence of hiatuses due to long intervals of erosion or non-deposition.

To investigate the possible effects of current velocity on the particle size distribution within the silt size range, trigger core-top samples were analyzed and interpreted in order to infer the depositional conditions on the east flank of the Vema Channel. The mean within the silt size range demonstrates a relationship which may be interpreted in view of the hydrography of the channel (Fig.12). The silt mean becomes finer with increasing water depth on the east flank of the channel, with the finest mean particle diameter in core 61. This core lies at a depth of 4181 m, which is near the benthic thermocline (Fig.5) but below the 1.2°C isotherm, which intersects the east flank of the channel at ~4000 m and has been interpreted to correspond approximately with the NADW/AABW transition zone (Johnson et al., 1976). In the absence of direct current observations, it is difficult to specify an appropriate level of no motion between north-flowing AABW and south-flowing NADW on the east flank of the channel. One might anticipate that the transition zone between opposite-flowing bottom currents would be the site of bottom current deposition, due to the loss of transport competence

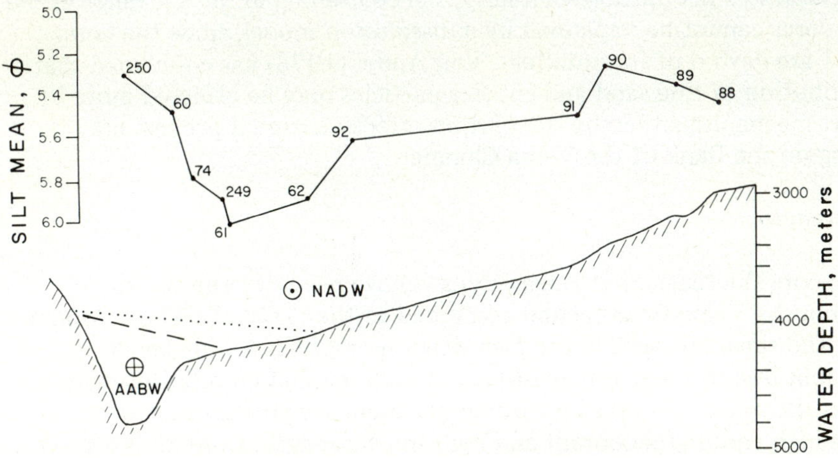

Fig.12. Mean particle diameter of silt size fraction in trigger core tops from the east flank of the Vema Channel. Dashed line sloping eastward identifies the position of the maximum gradient in the benthic thermocline (Fig.5). Dotted line identifies the 1.2°C potential temperature isotherm, which may correspond approximately with the upper limit of AABW (Johnson et al., 1976). Core locations are shown relative to geographic position on the east flank of the channel. Exact locations are indicated in Table I. Note that the finest mean silt particle size (in core 61) occurs where the benthic thermocline intersects the east flank of the channel, perhaps suggesting preferential deposition near a level of no motion in the AABW/NADW transition zone (see text for discussion).

caused by the edge effect of opposing currents (Davies and Laughton, 1972). The relatively fine particle size in core 61 (Fig. 12) and the thick Brunhes section in the same core (Fig.12) would support this interpretation, and would suggest that the level of no motion between NADW and AABW corresponds more closely with the benthic thermocline than with the 1.2°C isotherm. On the other hand, carbonate dissolution characteristics of the surface sediments (see previous section) indicate that the foraminiferal lysocline lies between cores 92 and 62 at ~4000 m, and that AABW may therefore extend to shallower depths than the benthic thermocline. Closely spaced direct current measurements will be required to resolve this uncertainty.

The cores deeper than core 61 are clearly beneath AABW, and the silt mean in these cores is significantly coarser than in core 61 (Fig. 12). The coarsening of the mean particle diameter in the silt range within parts of the Vema Channel is consistent with the presence of relatively fast-flowing AABW which has eroded or winnowed finer particles, leaving a coarse lag deposit.

An alternative interpretation for the relationship between decreasing particle size with increasing water depth could be increased dissolution at greater depth which produces a finer-grained sediment. However, the particle size analyses in this study included only the silt size range, which would become coarser due to the input of fragmented fine sand-size particles with greater carbonate dissolution. Similarly, the coarsening in the silt range in the deepest cores cannot be explained by a dissolution model, since the cores below 61 are devoid of foraminifera. Van Andel (1973) has concluded that the distribution of fine sand and silt size particles may be affected more by transport mechanisms than by dissolution effects; a similar process may be occurring on the flank of the Vema Channel.

Shear strength

Down-core fluctuations in shear strength have been determined for all cores from the Vema Channel and compared to the % $CaCO_3$. The carbonate content and shear strength in the four deepest cores show only small fluctuations due to the low magnitudes of each, so that correlations may not be tested. However, in cores 88 through 92 there is a strong correlation between high carbonate content and high shear strength. Core 88 is a good example of these relationships (Fig.13). The strong positive correlation may be the result of higher shear strengths due to greater abundance of carbonate particles (Bennett, 1972), or due to incipient carbonate cementation taking place during intervals of high carbonate deposition (Kelly et al., 1974).

The consistent positive correlation between carbonate content and shear strength in cores on the east flank of the Vema Channel leads to two interesting interpretations. First, the sediment deposited during interglacial stages, identified as $\delta\ ^{18}O$ minima (Fig.13), has a much higher strength. This

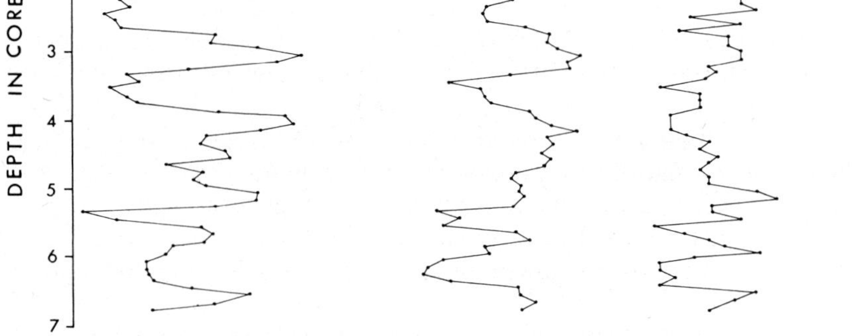

Fig.13. Plot of $\delta\ ^{18}O$ values (after Peters, 1976), percentage $CaCO_3$, and shear strength of sediment in core 88. Lower values of $\delta^{18}O$ are interpreted as periods of decreased ice volume (i.e., interglacial stages), and correlate with high $CaCO_3$ and high shear strength.

presumably would require higher current velocities before erosion may take place. Secondly, since the percentage of $CaCO_3$ has a strong depth control due to dissolution effects there may be some critical depth at which erosion may be enhanced due simply to decreasing calcareous material, without necessarily requiring higher bottom current velocities. This relationship might be important in interpreting the relative magnitudes of shallow and deep bottom currents. Lack of scour by bottom currents in relatively shallower regions does not necessarily infer slower velocities, since more cohesive sediments may be present in the shallower areas.

Dissolution indices

In order to evaluate the extent to which the Late Pleistocene carbonate cycles in cores 88 through 92 (Fig.8) may be dissolution controlled, a semi-quantitative measure of dissolution was obtained for three of the cores in the profile (Fig.14). From samples taken at 10- to 20-cm intervals in each core, foraminiferal assemblages were prepared consisting of the size fraction coarser than 250 μm. Between 300 and 500 individual specimens in each sample were counted, and designated as either fragments or whole tests. The ratio, P_f/P_w, is a semi-quantitative index of dissolution.

A comparison of the dissolution index with the carbonate content in the three cores (Fig.14) reveals several interesting features:

(1) Dissolution is of increasing importance at increasing water depth. In the shallowest core (No. 88), down-core variations in the dissolution index

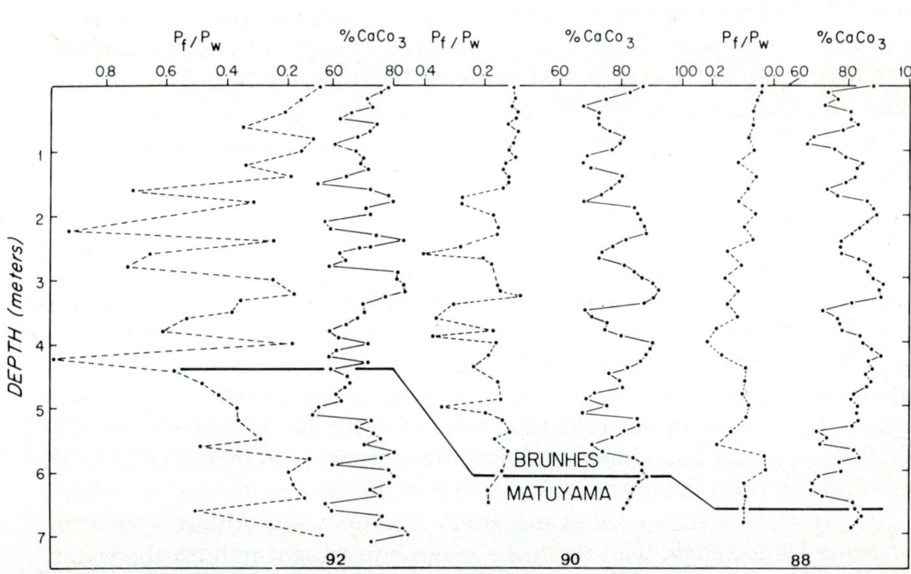

Fig.14. Foraminiferal dissolution index and $CaCO_3$ in three of the Vema Channel cores (88, 90, 92). The dissolution index, P_f/P_w, is the ratio of fragments to whole tests in counts of 300 and 500 specimens per sample.

P_f/P_w are minimal and probably not significant. In the deepest of the three cores (No. 92) the index varies between ~0.1 and >1.0. This interpretation is futher supported by the fact that the average Brunhes sedimentation rate decreases systematically with increasing water depth, and the mean carbonate content decreases from >80% in core 88 to ~70% in core 92. Consequently it appears that differential carbonate preservation with depth controls the varying average sedimentation rates.

(2) In core 92, and in the lower two-thirds of core 90, there is a strong positive correlation between low carbonate content and a high dissolution index. Since isotopic analysis indicates that carbonate maxima correspond with $\delta\ ^{18}O$ minima in shallower cores along this profile (Fig.13; Peters, 1976), then if the carbonate curves in Fig.14 can be reliably correlated from core to core, one would interpret the dissolution events to correspond to glacial episodes. Berger (1968) and Gardner (1975) previously suggested that glacial bottom water extended to shallower depths in the Atlantic, and our evidence supports this interpretation.

Antarctic diatoms

Displaced high-latitude diatoms have been used for some time to describe the path of spreading of AABW (Burckle and Biscaye, 1971; Burckle et al., 1973; Kolla et al., 1974; Connary and Ewing, 1974; Burckle et al., 1975). Basically the assumptions are as follows: Antarctic diatoms become entrained in the newly-formed AABW; as this water mass spreads northward, the diatoms are transported and eventually deposited on the sea floor. The presence of these forms at the sediment—water interface may therefore serve as a tracer of the path of spreading of AABW. Using this method, Burckle et al. (1975) found Antarctic diatoms as far north as 30°N in the Atlantic, while Booth and Burckle (1976) traced these forms into the equatorial regions of the Pacific. It should be pointed out that this method yields data that are in close agreement with physical oceanographic data on the path of spreading of AABW.

The study reported here permits us to determine the temporal distribution of Antarctic diatoms from the suite of "Chain" 115 cores in an area which may have been affected by vertical migrations of AABW during the Pleistocene. Preliminary work by one of us (LHB) has uncovered Antarctic-derived diatoms in Pleistocene sediments of the Brazil Basin and the central equatorial Pacific. A down-core study in a region such as the Vema Channel may therefore help to show: (1) the time of initiation of AABW; and (2) whether increased abundance of Antarctic diatoms (and increased bottom water production) are more characteristic of glacial or interglacial modes.

The principal species used in this study is the pennate Antarctic diatom *Nitzschia kerguelensis.* This form is a major constituent in both the water column and in the underlying sediments. It is found abundantly from the ice margin to the north of the Antarctic Convergence. In addition to *N. kerguelensis,* other forms may be useful tracers of the AABW. *Coscinodiscus lentiginosus* is also an abundant form and has been traced to the northern edges of the Argentine Basin. *Eucampia balaustium* is a heavily silicified near-ice form. Its presence in the sediments of the Argentine Basin is an indication of the transporting capacity of AABW.

Samples were taken at 25-cm intervals from six of the cores ("Chain" 115-59, 60, 61, 62, 88, 89, 92). Of these, only core 61 proved to contain a sizeable Antarctic diatom component. The others contained Antarctic diatoms in the surface and near-surface sediments, but were barren of diatoms, or had single diatom valves in a few down-core samples.

Core 61 had occurrences of Antarctic diatoms down to a depth of 750 cm. Not only was *N. kerguelensis* present, but there were also common occurrences of such forms as *C. lentiginosus* and *E. balaustium*. At 528 and at 378 cm, a single valve of *Hemidiscus karstenii* was observed. This form disappeared in southern ocean sediments at approximately 200,000 years B.P. (Abbott, 1972), and its presence here may help put some constraints on the age of the core. Although diatoms were observed to a depth of 750 cm in core 61, they are common only to a depth of 425 cm. Most of the diatoms and the major fluctuations in diatom abundance occur above this level. Interestingly, no diatoms were observed in the near-surface sample (3 cm) from this core.

Core 59 contained significant numbers of Antarctic diatoms in the upper 25 cm. Below that, single diatom values were observed at a few levels. Both cores 60 and 62 were barren of Antarctic-derived diatoms. Core 89, however, contained single valves of diatoms at several different levels. An interesting observation in this core is that opal phytoliths tend to be more common during carbonate highs. Core 92 contains diatoms down to the 25-cm level, with a single valve occurring at the 175-cm level.

DISCUSSION

Correspondence between surface sediment parameters and water properties

There are at least three sources of uncertainty in reliably establishing a one-to-one correspondence between diagnostic parameters of the surface sediment within the Vema Channel and characteristic properties of the overlying water:

(1) Water mass characteristics were not measured directly at each coring site. Extensive geochemical data ("Geosecs" Station 59) and hydrographic observations (Johnson et al., 1976) have been obtained in the axis of the channel to the west, but these measurements must be extrapolated laterally along isopycnal surfaces to infer water properties at the coring sites on the east flank of the channel.

(2) There are insufficient direct current measurements to specify flow characteristics in the NADW/AABW transition zone, and to establish with any degree of certainty the level of no motion.

(3) The coring process itself is of unknown reliability in obtaining representative samples of the upper sediment layers.

With the above limitations in mind, our data enable us to infer the following relations between observed sediment parameters and bottom water properties:

The abundance and nature of the carbonate component of the sediment is perhaps the most diagnostic of the sediment parameters. The foraminiferal lysocline is well defined at ~4000 m (between cores 92 and 62) on the east flank of the channel, and corresponds with hydrographic properties (θ = 1.2°C; σ_4 = 46.00) which may be appropriate choices for identifying the level of no motion in the NADW/AABW transition zone (Johnson et al., 1976). The CCD at 4250 m (between cores 61 and 60) corresponds to the position of the sharp benthic thermocline and maximum gradient in light scattering (Fig.5). Below this depth the near-bottom water is near-adiabatic, of near-neutral stability, and characterized by intense turbulent mixing (Johnson et al., 1976). The pteropod compensation depth at ~3200 m does not correspond to any known gradient in water characteristics, and therefore probably reflects an equilibrium situation between the production rate of aragonitic skeletal material and the rate of aragonite dissolution at and immediately below the sediment/water interface.

The mean grain size of the silt size fraction in core-top samples from the east flank of the channel may be a reliable indicator of relative current intensity. Abnormally high rates of deposition of relatively fine material are present in core 61, near the position of the benthic thermocline, with coarser sediment and slower rates of deposition (or perhaps erosion) at cores on either side (see Figs.10 and 12). Preferential deposition would be expected within a zone of convergence in the AABW/NADW transition zone on the east flank, since cross-contour flow due to advection in the bottom-boundary Ekman layer has been documented for this region of the Vema Channel (Johnson et al., 1976). The direction of this cross-contour component would be toward the east (upslope) for northward-flowing water, and toward the west (downslope) for southward-flowing water. Preferential deposition at core 61 suggests that this depth range (~4250 m) may mark such a zone of convergence. Direct current measurements would be required to verify this interpretation.

Antarctic diatoms as paleo-oceanographic indicators

Antarctic diatoms may or may not be indicative of the presence of AABW, depending probably on the near-bottom current velocity and the nature of the sediment/water interface. In the presence of a hard substrate (e.g., a manganese pavement) and/or relatively high current velocities, Antarctic diatoms may remain entrained in the near-bottom water and not be recorded in the depositional record. It appears, therefore, that the presence of these forms in sediments *is* clearly diagnostic of the presence of AABW, but their absence does *not* necessarily indicate the absence of AABW.

This discussion will be concerned primarily with Antarctic diatoms in core 61, since this is the only site which had a sizeable diatom component throughout much of the core. Fig.9 is a plot of the relative number of diatoms per gram of sediment in core 61. A number of observations can be made from this diagram. No Antarctic diatoms occur in the sediments below 720 cm, and there is a striking contrast in diatom abundance at about 425 cm with low counts below this level and higher counts above it. There is considerable fluctuation in diatom abundance above the 425-cm level (Fig.9).

How might these data be interpreted? Can we argue that the first appearance of the Antarctic diatoms in core 61 marks the initiation of AABW flow through the Vema Channel? Zimmerman et al. (1976) have recently presented evidence that the Vema Channel was not important as a passageway for the AABW until Pliocene times. Basing their interpretations on terrigenous sediment distribution, these authors believe that the Hunter Channel (Burckle and Biscaye, 1971) served as the principal passage for bottom water flow through the Rio Grande Rise during much of the Tertiary.

The data of Zimmerman et al. (1976) are given partial support from the record of displaced Antarctic diatoms in the Vema Channel. No diatoms are found in Vema Channel sediments of pre-Brunhes age. In core 61 we observe the first appearance of Antarctic diatoms within the Brunhes. The location of this core within the benthic thermocline on the flank of the Vema Channel suggests that it may be useful for recording volumetric changes in the AABW. Thus, the lack of Antarctic diatoms in the lower part of core 61 cannot be used to argue that no Antarctic Bottom Water was passing through the Vema Channel. Rather, it probably means that the upper level of the bottom water was below the site of core 61. The first occurrence of Antarctic diatoms at the 725 cm level would mark an increase in AABW transport in the channel, and the sharp increase in diatom abundance at the 425 level perhaps marks a further increase. Fluctuations in the abundance of diatoms above the 425 cm level may represent changes in volume of AABW going through the channel.

A second question to be considered is the precise age of these various events in core 61. The paleomagnetic stratigraphy and the presence of *Hemidiscus karstenii* provides us with the only clues. Abbott (1972) and the personal notes of one of us (Burckle) give an age of approximately 200,000 years for the last occurrence of this species. Its presence at 525 cm and 328 cm in core 61 places that part of the core below 328 cm in the lower to middle part of the *Coscinodiscus lentiginosus* zone of McCollum (1975). In the paleomagnetic stratigraphy this is approximately Early to Middle Brunhes.

Carbonate cycles and winnowing cycles

The correspondence between present hydrography and silt mean particle size in core top samples (Fig.12) may be used to predict the effect of pulses of AABW within the Vema Channel. If an increase in the volume of AABW results in a vertical migration of the NADW/AABW interface up the east flank of the channel, the result will be (1) a decrease in mean particle size at sites formerly above the interface; and (2) a coarsening of mean silt particle size at locations formerly near the interface but now under AABW. Pulses of AABW will also be recorded by fluctuations of mean particle diameter in cores *within* present AABW if the increased volume of AABW also resulted in increased bottom current velocity. However, such an increase in bottom current velocity within the constriction of the channel may result in total scour of sediment from the channel. Therefore, the best record of the increased velocity of AABW may be in cores from near the upper boundary of the flow. Paradoxically, the increased velocity may be recorded by a *decrease* in mean particle diameter of silt-sized particles at some depth intervals.

The two cores immediately shallower than the present upper level of AABW (as interpreted by the geological evidence in core 61) are cores 62 and 92. Core 62 is in the best location for testing the above model, since AABW may well have migrated vertically 100 m, necessary to have influenced the particle size distribution at that site. However, the top of core 62 appears to have not reliably recovered the near-surface sediment (see discussion of magnetic stratigraphy), and therefore the late Brunhes sediment may be missing from this core. Therefore, core 92 has been selected to test for pulses of AABW during the Late pleistocene. In order to influence the sedimentation pattern in core 92, AABW must have increased sufficiently in volume to cause a rise of 300 m above its present location. It cannot be determined if this is a reasonable increase of AABW without additional data from cores at the critical depth near core 62 which could be used to trace pulses of AABW up the east flank of the channel.

Particle size analysis of the silt fraction in core 92 reveals that increases in mean particle diameter are coupled with more positive skewness and with higher percentages of $CaCO_3$ (Fig.15). The linear correlation coefficient between mean silt particle diameter (expressed in phi units) and % $CaCO_3$ is −0.24, which corresponds to a significance level of >95%. Increases in % $CaCO_3$ are correlated with interglacial intervals, as interpreted from $\delta\ ^{18}O$ data from two of the Vema Channel cores (Fig.13; Peters, 1976). Therefore, glacial stages (i.e., periods of low $CaCO_3$) are characterized by finer mean diameter of silt particles and more negative skewness in the particle size distribution. Each of the textural parameters suggests that bottom currents were flowing with a lower velocity at a depth of 3934 m during glacial stages.

The reduction in apparent current velocity during glacial periods on the east flank of the Vema Channel may be attributed to either (1) reduction in current velocity of southward-flowing NADW, or (2) an increased volume of northward-flowing AABW, such that the boundary between the two opposing bottom currents shallows by approximately 300 m. With a better coverage of cores in the region of the present upper level of AABW which may be correlated on the basis of glacial/interglacial stages, the problem may be solvable. However, if more detailed $\delta\ ^{18}O$ and nannofossil stratigraphy becomes available for the deeper cores, the answer may be determined.

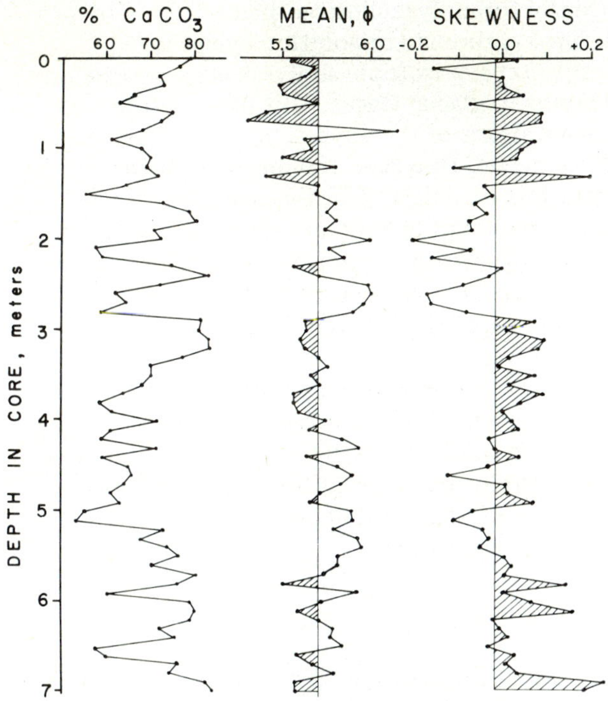

Fig.15. Plot of percentage $CaCO_3$, mean particle diameter in silt size range, and skewness of silt-sized particle distribution in core CH 115-6-92. Vertical line in mean and skewness data represents the mean of all data. The cross-hatched areas designate intervals of inferred higher bottom current velocity. These intervals correlate with interglacial periods, identified by high % $CaCO_3$ and $\delta\ ^{18}O$ minima.

SUMMARY

(1) Surface sediment samples from the east flank of the Vema Channel reflect the present-day hydrography near the AABW/NADW transition zone, and suggest the presence of at least two important changes in water characteristics within this zone. The foraminiferal lysocline, which intersects the east flank at ~4000 m, corresponds approximately with $\theta = 1.2°C$ and the top of the benthic thermocline, and may be an appropriate choice for a level of no motion within the transition zone (Johnson et al., 1976). The CCD, which intersects the east flank at ~4250 m, corresponds with the maximum gradient in the benthic thermocline, and the rapid deposition of fine-grained sediment. Such preferential deposition would be expected in a zone of convergence on the east flank between northward-flowing water and overlying southward-flowing water, due to cross-contour Ekman transport in the bottom boundary layer. We have insufficient direct current measurements to verify whether either of these two geological horizons (the foram lysocline and the CCD) does in fact correspond to the level of no motion.

(2) Piston cores from above the foram lysocline on the east flank (depth 2900—4000 m) show well-defined carbonate dissolution cycles with periodicities of ~100,000 years. Oxygen isotopic analyses indicate that maximum dissolution corresponds to glacial stages. This interpretation is consistent with previous interpretations of the abyssal circulation in the central Atlantic during the Pleistocene. However, the mechanism for producing this cyclic sedimentation is unclear. The absence of Antarctic diatoms within the dissolution cycles, and the apparently complete Brunhes depositional record in each core, suggests that extensive vertical migrations of corrosive AABW may not necessarily have occurred. Fluctuations in the characteristics and/or production rate of NADW may be a more likely explanation.

(3) Particle size data in the silt-size range and the presence of displaced Antarctic diatoms reflect the present-day hydrography, and may therefore be useful parameters for identifying fluctuations in bottom water characteristics. However, initial data on grain size fluctuations in core 92 and displaced Antarctic diatoms in core 61 suggest that such pulses may have been of only minimal extent during the Brunhes.

ACKNOWLEDGEMENTS

Shipboard operations in the Vema Channel and Rio Grande Rise region ("Chain" 115, Leg 6) were supported under NSF Grant DES74-01744 and ONR Contract No. N00014-74-C0262. Paleomagnetic stratigraphy, grain size analysis, and shear strength determination of core samples were performed at laboratory facilities of the University of Rhode Island. We thank N.D. Watkins, T.C. Huang, and W.E. Kelly (of U.R.I.) for encouragement and for the use of laboratory equipment, and B.B. Ellwood for extensive computer programming assistance. MTL received support under NSF Grant DES75-13839 for Doctoral Dissertation Research. LHB is supported under NSF Grant No. IDO75-19627. R. McGirr and C. Peters performed the carbonate analysis, and A. Riley obtained the foraminiferal dissolution indices. We thank G.P. Lohmann, E. Laine, and J. Thiede for their constructive criticisms of the manuscript.

REFERENCES

Abbott, W.H., 1972. Vertical and Lateral Patterns of Diatomaceous Ooze found between Australia and Antarctica. Thesis, Univ. South Carolina, Columbia, S.C., pp.1—136.

Bennett, R.H., 1972. Paleotemperature and cohesion in *Globigerina* ooze sediment cores from the Caribbean Sea. Nature, 240: 114—116.

Berger, W.H., 1968. Planktonic foraminifera: selective solution and paleoclimatic interpretation. Deep-Sea Res., 15: 31—43.

Berggren, W.A. and Hollister, C.D., 1974. Paleogeography, paleobiogeography, and the history of circulation in the Atlantic Ocean. Soc. Econ. Paleontol. Mineral. Spec. Publ., 20: 126—186.

Biscaye, P.E. and Eittreim, S.L., 1974. Variations in benthic boundary layer phenomena: nepheloid layer in the North American Basin. In: R.J. Gibbs (Editor), Suspended Solids in Water. Plenum, New York, N.Y., pp.227—260.
Booth, J.D. and Burckle, L.H., 1976. Displaced Antarctic diatoms in the southwestern and central Pacific. Pac. Geol., in press.
Burckle, L.H., 1976. Use of diatoms as tracers of the northward flow of Antarctic Bottom Water. Physis, in press.
Burckle, L.H. and Biscaye, P.E., 1971. Sediment transport by Antarctic Bottom Water through the eastern Rio Grande Rise. Geol. Soc. Am., Abstr. Progr., 3: 518—519 (abstract).
Burckle, L.H., Abbott, W.H. and Maloney, J., 1973. Sediment transport by Antarctic Bottom Water in the southeast Indian Ocean. EOS Trans. Am. Geophys. Union, 54: 336 (abstract).
Burckle, L.H., Poferl-Cooke, K. and Maloney, J., 1975. Southward flowing Antarctic Bottom Water off Cape Hatteras — evidence from displaced Antarctic diatoms. Geol. Soc. Am., Abstr. Progr., 7: 34 (abstract).
CLIMAP, 1976. The surface of the ice age earth. Science, 191: 1131—1137.
Connary, S.D., 1972. Investigations of Walvis Ridge and Environs. Thesis, Columbia Univ., New York, N.Y., 228 pp.
Connary, S.D. and Ewing, M., 1974. Penetration of Antarctic Bottom Water from the Cape Basin into the Angola Basin. J. Geophys. Res., 79: 463—469.
Davies, T.A. and Laughton, A.S., 1972. Sedimentary processes in the North Atlantic. In: A.S. Laughton, W.A. Berggren et al. (Editors), Initial Reports of the Deep Sea Drilling Project, 12. U.S. Government Printing Office, Washington D.C., pp.905—934.
Diester-Haass, L., 1976. Influence of deep-oceanic currents on calcareous sands off Brazil. In: Proc. IX Congres International de Sedimentologie, Nice, Thème 8, pp. 25—2
Gardner, J.V., 1975. Late Pleistocene carbonate dissolution cycles in the eastern equatorial Atlantic. In: Dissolution of Deep-Sea Carbonates. Cushman Found. Foramini Res., Spec. Publ., 13: 129—141.
Gordon, A.L., 1972. Spreading of Antarctic bottom waters, II. In: A.L. Gordon (Editor), Studies in Physical Oceanography — A Tribute to G. Wüst on His 80th Birthday. Gordon and Breach, New York, N.Y., pp.1—18.
Hobart, M.A., Bunce, E.T. and Sclater, J.G., 1975. Bottom water flow through the Kane Gap, Sierra Leone Rise, Atlantic Ocean. J. Geophys. Res., 80: 5083—5088.
Hollister, C.D. and Heezen, B.C., 1972. Geologic effects of ocean bottom currents. In: A.L. Gordon (Editor), Studies in Physical Oceanography — A Tribute to G. Wüst on His 80th Birthday. Gordon and Breach, New York, N.Y., pp.37—66.
Hollister, C.D., Johnson, D.A. and Lonsdale, P.F., 1974. Current-controlled abyssal sedimentation: Samoan Passage, equatorial west Pacific. J. Geol., 83: 275—300.
Huang, T.C. and Watkins, N.D., 1977. Contrasts between the Brunhes and Matuyama sedimentary records of bottom water activity in the South Pacific. Mar. Geol., 23: 113—132.
Huang, T.C., Watkins, N.D. and Shaw, D.M., 1975. Atmospherically transported volcanic glass in deep-sea sediments: development of a separation and counting technique. Deep-Sea Res., 22: 185—196.
Johnson, D.A., 1972. Ocean-floor erosion in the equatorial Pacific. Geol. Soc. Am. Bull., 83: 3121—3144.
Johnson, D.A., 1974a. Deep Pacific circulation: intensification during the early Cenozoic. Mar. Geol., 17: 71—78.
Johnson, D.A., 1974b. Initial Cruise Report, Chain 115, Leg 6. Woods Hole Oceanogr. Inst., Tech. Rep., 74-39: 51 pp.
Johnson, D.A. and Sullivan, L.G., 1976. Light scattering observations in the Vema Channel and on the Rio Grande Rise, Chain 115, Leg 6. Woods Hole Oceanogr. Inst., Tech. Rep., 76-2, 24 pp.
Johnson, D.A., McDowell, S.E. and Von Herzen, R.P., 1975. Hydrographic and abyssal

temperature data from the Vema Channel and Rio Grande Rise, Chain 115, Leg 6. Woods Hole Oceanogr. Inst., Tech. Rep., 75-4: 41 pp.

Johnson, D.A., McDowell, S.E., Sullivan, L.G. and Biscaye, P.E., 1976. Abyssal hydrography, nephelometry, currents, and benthic boundary layer structure in the Vema Channel. J. Geophys. Res., in press.

Jones, E.J.W., Ewing, M., Ewing, J.I. and Eittreim, S.L., 1970. Influences of Norwegian Sea overflow water on sedimentation in the northern North Atlantic and Labrador Sea. J. Geophys. Res., 75: 1655—1680.

Kelly, W.E., Nacci, V.A., Wang, M.C. and Demars, K.R., 1974. Carbonate cementation in deep-ocean sediments. J. Geotech. Eng. Div., ASCE, 95(SM5): 383—386.

Kennett, J.P., Houtz, R.E. et al., 1975. Initial Reports of the Deep Sea Drilling Project, 29. U.S. Government Printing Office, Washington, D.C.

Kolla, V., Burckle, L.H. and Booth, J.D., 1974. Sediment transport of Antarctic Bottom Water in the western Indian Ocean. EOS Trans. Am. Geophys. Union, 55: 312 (abstract).

Le Pichon, X., Eittreim, S.L., and Ludwig, W.J., 1971a. Sediment transport and distribution in the Argentine Basin, 1. Antarctic bottom current passage through the Falkland Fracture Zone. In: L.H. Ahrens, F. Press, S.K. Runcorn and H.C. Urey (Editors), Physics and Chemistry of the Earth, 8. Pergamon, New York, N.Y., pp.3—28.

Le Pichon, X., Ewing, M. and Truchan, M., 1971b. Sediment transport and distribution in the Argentine Basin, 2. Antarctic bottom current passage into the Brazil Basin. In: L.H. Ahrens, F. Press. S.K. Runcorn and H.C. Urey (Editors), Physics and Chemistry of the Earth, 8. Pergamon, New York, N.Y., pp.31—48.

Lonardi, A.G. and Ewing, M., 1971. Sediment transport and distribution in the Argentine Basin, 4. Bathymetry of the continental margin, Argentine Basin and other related provinces. Canyons and sources of sediments. In: L.H. Ahrens, F. Press, S.K. Runcorn and H.C. Urey (Editors), Physics and Chemistry of the Earth, 8. Pergamon, New York, N.Y., pp.79—121.

Lonsdale, P. and Southard, J.B., 1974. Experimental erosion of North Pacific red clay. Mar. Geol., 17: M51—M60.

McCave, I.N. and Jarvis, J., 1973. Use of the Model T Coulter Counter in size analysis of fine to coarse sand. Sedimentology, 20: 305—315.

McCollum, D., 1975. Diatom stratigraphy of the southern ocean. In: D. E. Hayes, L.A. Frakes et al. (Editors), Initial Reports of the Deep Sea Drilling Project, 28. U.S. Government Printing Office, Washington, D.C., pp.515—571.

Melguen, M. and Thiede, J., 1974. Facies distribution and dissolution depths of surface sediment components from the Vema Channel and the Rio Grande Rise (southwest Atlantic Ocean). Mar. Geol., 17: 341—353.

Peters, C., 1976. Oxygen isotopic analysis of two cores from the Vema Channel: an evaluation of the method and results. Woods Hole Oceanogr. Inst., Tech. Rep., 76-10: 1—29.

Reid, J.L., Nowlin, W.D., McLellan, H.J. and Patzert, W.C., 1973. Deep and bottom flow in the southwest Atlantic. EOS Trans. Am. Geophys. Union, 54: 313 (abstract).

Reid, J.L. and Lonsdale, P.F., 1974. On the flow of water through the Samoan Passage. J. Phys. Oceanogr., 4: 58—73.

Schick, G.B., Isaacs, J.D. and Sessions, M.H., 1968. Autonomous instruments in oceanographic research. In: Marine Science Instrumentation, 4. Plenum, New York, N.Y., pp.203—230.

Schrader, H.J., 1974. Proposal for a standardized method of cleaning diatom-bearing deep-sea and land-exposed marine sediments. Nova Hedwigia, Beih., 45: 403—409.

Southard, J.B., Young, R.A. and Hollister, C.D., 1971. Experimental erosion of fine abyssal sediment. J. Geophys. Res., 76: 5903—5909.

Swift, D.J.P., Schubel, J.R. and Sheldon, R.E., 1972. Size analysis of fine-grained suspended sediments: a review. J. Sediment. Petrol., 42: 122—134.

Van Andel, Tj. H., 1973. Texture and dispersal of sediments in the Panama Basin. J. Geol.

81: 434—457.
Walker, P.H., Woodyear, K.D. and Hutka, J., 1974. Particle-size measurements by Coulter Counter of very small deposits and low suspended sediment concentration in streams. J. Sediment. Petrol., 44: 673—679.
Watkins, N.D. and Kennett, J.P., 1971. Antarctic Bottom Water: major change in velocity during the Late Cenozoic between Australia and Antarctica. Science, 173: 813—818.
Wright, W.R., 1970. Northward transport of Antarctic Bottom Water in the western Atlantic Ocean. Deep-Sea Res., 17: 367—371.
Wüst, G., 1957. Stromgeschwindigkeiten und Stromungen in den Tiefen des atlantischen Ozeans. Wiss. Ergeb. Dtsch. Atl. Exped. "Meteor" 1925—1927, 6 (part 2): 420 pp.
Zimmerman, H.B., McCoy, F.W. et al., 1976. Sediment lithofacies as indicators of the paleo-oceanographic environment in the south Atlantic Ocean. EOS Trans. Am. Geophys. Union, 57 (abstract).

Marine Geology, 23 (1977) 35—56

PALEOCURRENTS IN THE EASTERN CARIBBEAN: GEOLOGIC EVIDENCE AND IMPLICATIONS

TROY L. HOLCOMBE* and WILLARD S. MOORE**

U.S. Naval Oceanographic Office, Washington, D.C. (U.S.A.)

(Received April 28, 1976)

ABSTRACT

Holcombe, T.L. and Moore, W.S., 1977. Paleocurrents in the eastern Caribbean: geologic evidence and implications. Mar. Geol., 23: 35—56.

Late Cretaceous and Cenozoic sedimentary strata in the southern Venezuela Basin thin to less than 200 m across an east—west crescent-shaped depositional minimum which is 50—100 km wide and more than 400 km long. Drilling has revealed the presence of Santonian-to-late Paleocene and early Eocene-to-early Miocene unconformities along the axis of thinnest sediments. Sediments also thin over the crests of the Beata and Aves ridges, where unconformities in the late Cretaceous-to-Miocene time frame have been revealed or inferred by drilling. The relationship of sediments and structure on the Aves and Beata ridges suggests that these ridges were structural prominences for the duration of time represented by the unconformities. Apparently the Venezuela Basin unconformities are a record of long periods of nondeposition in a regime of sediment transport by currents. Renewed deposition across the unconformity in Miocene time correlates with renewed uplift and volcanism in Panama and the Lesser Antilles, which probably decreased water depth and cross-section to the point of restricting circulation. Renewed deposition in the Miocene also correlates with subsidence of the Aves Ridge, and this could have been another factor in the alteration of the current regime. We suggest that westward flowing currents swept the major crests of the Aves and Beata ridges and the unconformity area of the Venezuela Basin before closure of the Atlantic–Pacific marine connection through the Caribbean.

INTRODUCTION

One of the noteworthy features of the deep-sea sediment section beneath large areas of the eastern Caribbean is the uniformity of sediment thickness and reflective character, as revealed by seismic reflection data. The observed uniformity highlights anomalous areas such as the southern Venezuela Basin, where the sediment section thins appreciably. An early seismic reflection traverse of what was later revealed to be the western extremity of this thin-

---

*Present address: Naval Ocean Research and Development Activity, National Space Technology Laboratories, Bay Saint Louis, Mississippi 39520 (U.S.A.).
**Present address: Geology Department, University of South Carolina, Columbia, S.C. (U.S.A.).

sediment area showed contrasting sediment thickness on opposite sides of a broad, low rise. Taking note of this occurrence, Edgar et al. (1971) suggested that bottom currents may have been a factor in redistributing sediments. A drilling site survey further defined sediment patterns in the area, and D.S.D.P. drill holes 146/149, 29, and 150 were positioned to provide lithologic and time-stratigraphic control (Edgar et al., 1973a). It was determined that major unconformities in the late Cretaceous and Cenozoic sediment section coincide with the axis of sediment thinning. Unconformities of similar time frame were discovered at drill sites located atop the Beata and Aves ridges. In analyses of the evidence, Edgar et al. (1973b) and Holcombe et al. (1973) explained the absence of sediments as being the result of paleocurrents.

This paper presents a summary of the geologic evidence for paleocurrents in the Venezuela Basin, followed by a discussion of the nature of the current regime. The structurally positive elements of the eastern Caribbean, particularly the Beata and Aves ridges, are likewise considered in terms of evidence for currents and topographic effects on the current regime. Finally, the paper includes a discussion of implications for global water circulation and global tectonics. Fig.1 is a reference map of the eastern Caribbean showing the major physiographic features and the locations of the drill holes, seismic sections, and maps.

UNCONFORMITIES IN THE VENEZUELA BASIN—EVIDENCE FOR PALEOCURRENTS

Conspicuous breaks in the late Cretaceous and Cenozoic sedimentary record occur at two drilled sites in the Venezuela Basin (Edgar et al., 1973a). These hiatuses coincide with an arcuate, east—west, elongated belt where the sedimentary section thins to a minimum of less than 200 m. Elsewhere in the Venezuela Basin, the post-late Cretaceous strata are of fairly uniform 600—1000 m thickness.

The hiatuses are apparent in the biostratigraphic and lithologic summaries of Caribbean drill sites occupied by the Deep-Sea Drilling Project (Saunders et al., 1973). At site 150 (14°31′N, 69°22′W), centered on the axis of thin sediments, two major interruptions of sedimentation occur, one of Santonian—late Paleocene duration, and the other of early Eocene—early Miocene duration. A total of about 50 m.y. time is represented by the missing stratigraphic intervals at site 150. At site 146/149 (15°07′N, 69°22′W), located about 60 km north of site 150, outside the belt of thin sediments, no significant sediment breaks occur. At site 29 (14°47′N, 69°20′W), situated at the margin of the belt of thinning between sites 150 and 146/149, drilling did not penetrate beyond early Eocene strata, but a single interruption or condensation of the section, extending in time from latest Eocene to early Miocene, is observed. Late Turonian and Coniacian sediments immediately overlie igneous rock of basaltic composition at sites 146/149 and 150. Between these two sites, situated in a north—south line across the belt of minimum sediment thickness, the sediment section overlying the igneous horizon thins from 768 m at site 146/149 to 168 m at site 150 (Fig.2).

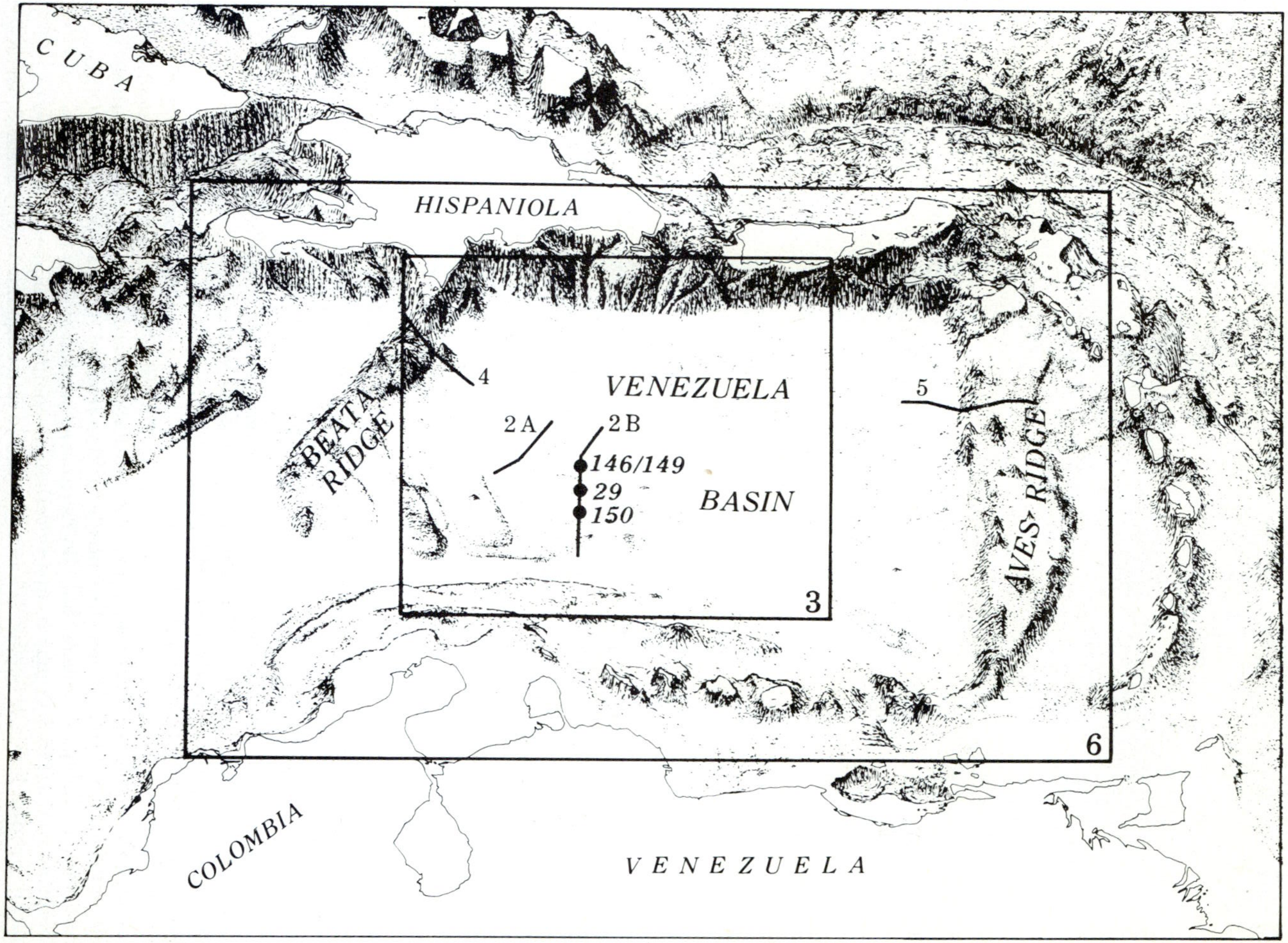

Fig.1. Physiographic sketch of the eastern Caribbean showing locations of sections in Figs.2A, 2B, 4, and 5, map boundaries for Figs.3 and 6, and locations of drill holes 146/149, 29, and 150. Physiography courtesy of R.N. Bergantino.

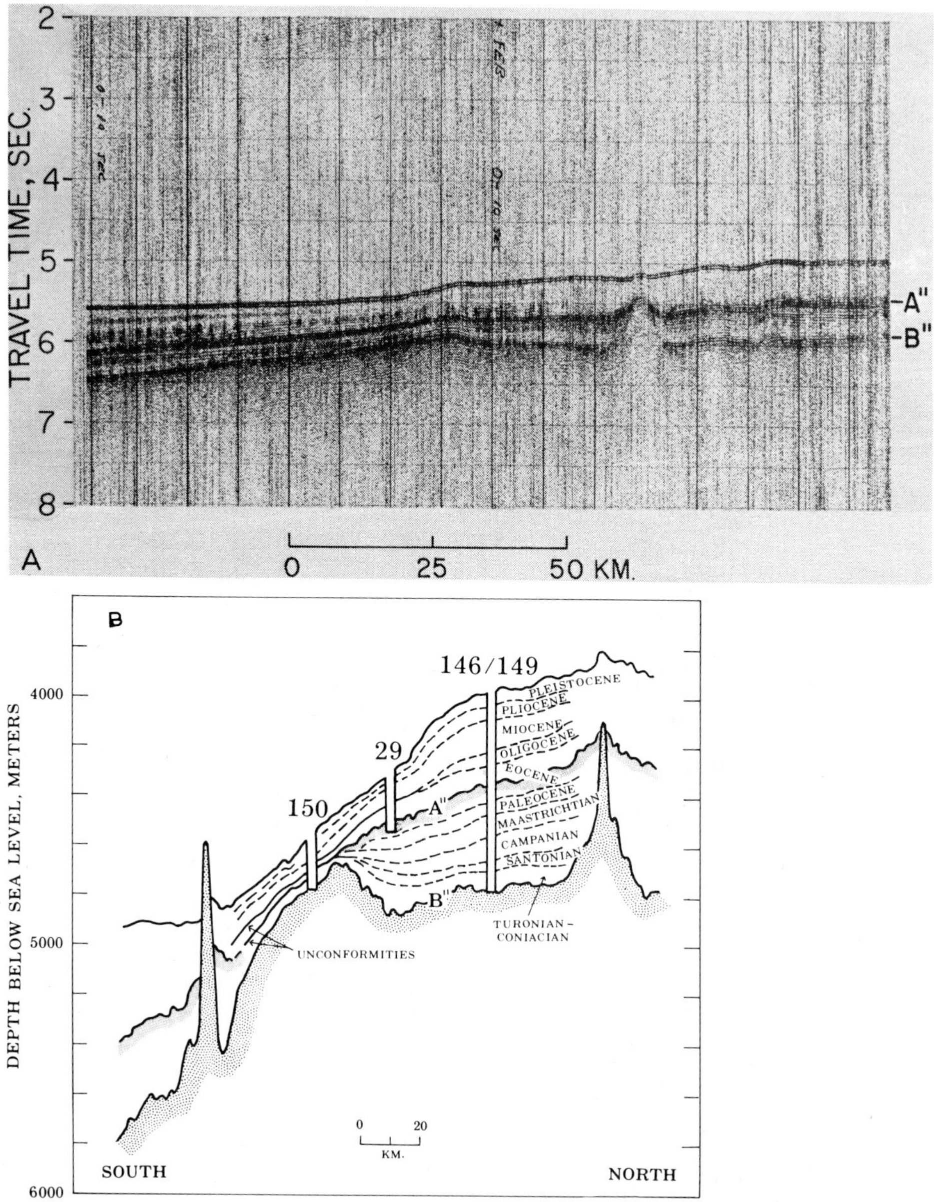

Fig.2A. Seismic reflection section *A-A'* across western end of thin sediments belt. Location shown in Fig.1. Horizons *A"* and *B"* are labeled. Note that sediment reaches minimum thickness left of broad structural high in center of section.
B. Geologic cross-section through DSDP drill holes across axis of sediment thinning in Venezuela Basin. Horizons *A"* and *B"* are labeled. Time-stratigraphic units, delineated by dashed lines, are inferred between drill holes. Note southward slope of unconformities and underlying horizons. Apparently nondeposition reached its greatest extent in Oligocene time. Thinning of Miocene and Plio-Pleistocene section at sites 150 and 29 as compared to site 146/149 may be due to carbonate compensation depth. Location of trackline shown in Fig.1. Section is based on seismic reflection line run by the U.S. Geological Survey (Silver et al., 1972). Depth corrections are based on sound speed data obtained at drill site 146/149 (Edgar et al., 1973a). Vertical exaggeration = 100.

The belt of sediment thinning is delineated by two prominent seismic reflection horizons which are believed to be roughly time-stratigraphic in this area (Saunders et al., 1973). The persistence and uniformity of these reflecting horizons (horizons A″ and B″) over the Venezuela Basin has been discussed (Ewing et al., 1967, 1968; Edgar et al., 1971). At the drill sites, horizon A″ coincides with the first occurrence of lithified sediments and chert horizons of early Eocene age. Horizon B″ coincides with the aforementioned occurrence of basalt beneath late Turonian—Coniacian sedimentary strata. The depositional patterns which define the extent of the sediment thinning have been described and discussed in some detail (Edgar et al., 1973b; Holcombe et al., 1973). Broadly speaking, the belt constitutes a "depositional scar" 50–100 km wide and more than 400 km in length (Fig.3). Its crescent shape is oriented so that it is trending east—west at its center, northeast toward its eastern end (14°30′N, 67°30′W) and northwest at the western extremity. At one point on the crescent (about longitude 68°W), thin sediments in the A″—B″ interval extend southeastward to the outer edge of the Basin.

The structural framework of the central Venezuela Basin is one of very broad, low-relief, northeast- and northwest-trending structural swells and troughs producing a roughly orthogonal fabric (Holcombe and Matthews, 1973; Matthews and Holcombe, 1974; Case and Holcombe 1975). The axis of the depositional minimum exhibits structural control; it parallels a major northeast-trending structural high along its eastern limb and a northwest-trending high along its western limb (Fig.3). The southeastward extension of thin sediments follows the northwest-trending structural high.

An ocean current of sufficient competence to impede or inhibit deposition apparently produced the axis of minimum deposition and, hence, the sediment hiatuses (Edgar et al., 1973b; Holcombe et al., 1973). Nondeposition due to scarcity of pelagic source sediments may be ruled out because it does not explain the total absence of strata representing about 50 m.y. of time at site 150, as compared to pelagic thicknesses observed elsewhere in the Venezuela Basin; even though the sediments at site 150 contain less biogenic calcium carbonate than those at site 146/149, one would expect at least a thin sedimentary section to accumulate below the carbonate compensation depth. Long periods of normal pelagic deposition followed by intervals of substantial erosion may be discounted. No beveling of strata is apparent; instead, the strata appear to converge. Also, no significant structural dislocation or channeling in the post-B″ sedimentary column is observed. The observations may be reasonably explained as depositional effects of a long-continued bottom-current regime, adjusted to a broad, low, structural relief which has retained its shape and stability since late Cretaceous time.

Stability of the current regime and basement structure is apparent, in that no significant lateral shifting of the current axis is evident in the sediment thickness patterns across the depositional minimum (Figs.2 and 3). A long-continued period of limited deposition is further verified by the occurrence of iron—manganese nodules, ranging up to five cm in diameter, in early

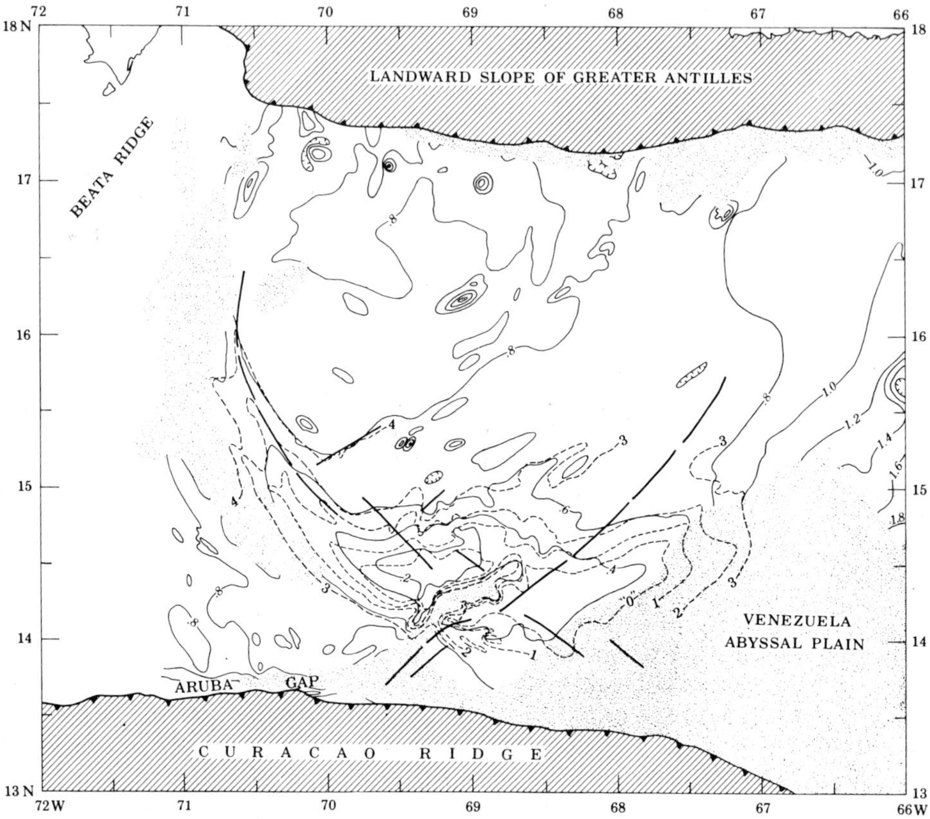

Fig.3. Sediment thickness patterns in Venezuela Basin. Turbidite plains shown by shaded areas. Solid isopachs show thickness of sedimentary strata above horizon *B″* in seconds of two-way travel time. Dashed isopachs show thickness of the *A″*-to-*B″* interval in meters × 100, assuming a sound speed of 2.47 km/sec based on velocity data obtained at drill-site 146/149 (Edgar et al., 1973a). The "0" thickness in the *A″*-*B″* interval seen in seismic reflection records is only apparent; at drill site 150 (Edgar et al., 1973a), sediments occur in the *A″*-to-*B″* interval, but the section is too thin to be resolved acoustically. Note the arcuate shape of the depositional minimum. The heavy lines show the broad, low, pre-Cenozoic structural highs which appear to have influenced the position of the currents. Note parallelism of western half of depositional minimum and northwest trending structural high and similar parallelism with northeast trending structural high at eastern end.

Miocene sediments immediately overlying the upper unconformity at drill site 150 (Edgar et al., 1973a). Manganese is thought to accumulate slowly (0.1—0.4 cm 10^{-6} yr^{-1}) and ubiquitously on the ocean floor (Bender et al., 1966; Krishnaswami et al., 1972; Moore, 1973). Occurrence of manganese nodules, however, is limited to areas of very slow sediment deposition, either below the carbonate compensation depth and far from land, or where bottom currents prevent or inhibit deposition (Heezen and Hollister, 1971; Horn et al., 1972). Although the nodules from site 150 are scattered through about four meters of vertical section near the base of one of the drilling cores (Edgar et al., 1973a), the core shows evidence of severe drilling disturbance, which would not seem to preclude their being concentrated at a single stratigraphic level, that of the unconformity. If they do occur enclosed within the early Miocene sediments above the hiatus, their level of occurrence could represent the level at which they were buried as the sedimentation rate began to increase.

At site 150 total nondeposition for substantial time periods has occurred, probably without substantial erosion, implying a sedimentation regime of transportation at the axis of the current. The grain size content of the pelagic section at drill site 146/149, outside the depositional minimum, bears on what size range of material was probably being transported. At site 149 (Edgar et al., 1973a), in the early Miocene-to-middle Eocene section, the silt and sand fraction ranges from 30—50% of the total content. The sand fraction varies from 0—10%. Percentages of sand and silt both increase downward in the section. At site 146 (ibid.), percentages of sand and silt are high (20% and 50%, respectively) at two levels, one in the early Eocene and the other in the Paleocene. Throughout most of the early Eocene-Santonian section, sand plus silt content is variable, ranging between 10 and 45% and increasing downward in the section; sand sized material is negligible through most of the Paleocene and Maastrichtian section, and 10% or less below the Maastrichtian. Elsewhere in the Caribbean the sand fraction is generally less than 10—15% and the sand plus silt fraction is generally less than 50%.

Since the current had available to it pelagic sediments of the same particle size range deposited in adjacent waters, the current at its axis must have flowed with sufficient speed to transport silt and probably fine sand, but without sufficient speed to erode sediments of this size range. A sandy silt interval occurs in the early Eocene of site 150, in the sediments separating the two unconformities (Edgar et al., 1973a). The size analysis shows an unusually high sand percentage (~40%), suggesting that at the time of deposition, either the current had decreased in speed to the extent that it was not competent to transport part of its suspended sand-sized load, or that current speed remained constant but a certain percentage of material in grain sizes too large to be transported were available to the current.

From a recent summary of experimental and theoretical suspension and traction curves (Hollister and Heezen, 1973), one may deduce that current speeds at the axis were probably on the order of 1—20 cm sec^{-1} at the bottom, considering the probable size range of material transported. At bottom speeds higher than 20 cm sec^{-1}, one might expect active erosion to have occurred, whereas at speeds of less than 1 cm sec^{-1}, deposition of the coarser fraction would take place. More detailed composition and grain size analyses of sediment cores from drill sites 146/149, 29 and 150 might yield more accurate information about probable current speeds.

PALEOCURRENTS OVER THE BEATA AND AVES RIDGES

A current of sufficient competence to account for nondeposition in the Venezuela Basin must have left a record of nondeposition on the Aves and Beata ridges, assuming that those ridges were in existence during late Cretaceous to Miocene time. It seems probable that the stability of the current effects observed in the Venezuela Basin could not have been maintained through late Cretaceous and early Tertiary time without the stable framework, relatively speaking, of the Beata Ridge, Aves Ridge, and Venezuela Basin.

Seismic reflection data strongly suggest that the Beata Ridge was in existence by latest Cretaceous time. Ewing, et al. (1967) noted that horizon B″, albeit structurally deformed, continues over the Beata Ridge, whereas horizon A″ appears to lap onto the upturned slopes of horizon B″, and they cited this as evidence that the Beata Ridge structures are probably post-B″ and pre-A″ in age. Cores containing shallow water mid-Eocene carbonates from the crest of the Ridge establish that the Ridge was formed by middle Eocene (pre-A″) time (Fox and Heezen, 1975). Manganese crusts 7—10 cm thick which have been dredged from the western scarp of the Beata Ridge (Fox et al., 1970) suggest an age on the order of 50—100 m.y., assuming a normal hydrogenous accumulation rate of manganese. Basaltic rocks dredged from the scarps of the Beata Ridge exhibit an advanced degree of weathering (ibid.).

Additional seismic reflection evidence suggests that the principal structural uplift of the Beata Ridge may have occurred no later than the early part of the post-B″, pre-A″ time span. The post-horizon-B″ sediment section appears to thin over structural highs and thicken in basins. Interval thicknesses in one basin appear to be independent of thicknesses in adjoining basins (Fig.4). The conformity of thickness patterns to the structure, implies pre-existence of the structure. This pattern is observed in the surface-to-A″ and A″-to-B″ intervals, both on the main ridge and the step-like fault blocks adjacent to the main ridge. However, some structures on the ridge, and a series of northwest-trending structures in the southwest part of the Venezuela Basin (Ewing et al., 1967; Roemer et al., 1973; Holcombe and Matthews, 1973) have clearly been subjected to more recent, probably Miocene or younger, structural displacements. Formation of the Beata Ridge possibly coincided in time with the emplacement of the horizon B″ sills. Occurrence of basaltic ashes in the strata above the sills is limited to the rocks of late-Turonian to Campanian age (Saunders et al., 1973), and they (ibid.) suggest that this represents the time interval during which the sills were deposited.

Formation of the Aves Ridge may have occurred in late Cretaceous time, prior to initiation of nondeposition in the southern Venezuela Basin. Granodiorites of probable late Cretaceous age obtained from escarpments near the southern end of the Aves Ridge suggest that emplacement of granitic plutons in the late Cretaceous formed the plateau of the ridge (Fox et al., 1971). Volcanic rocks from the higher ridges, seamounts, and pedestals of the Aves

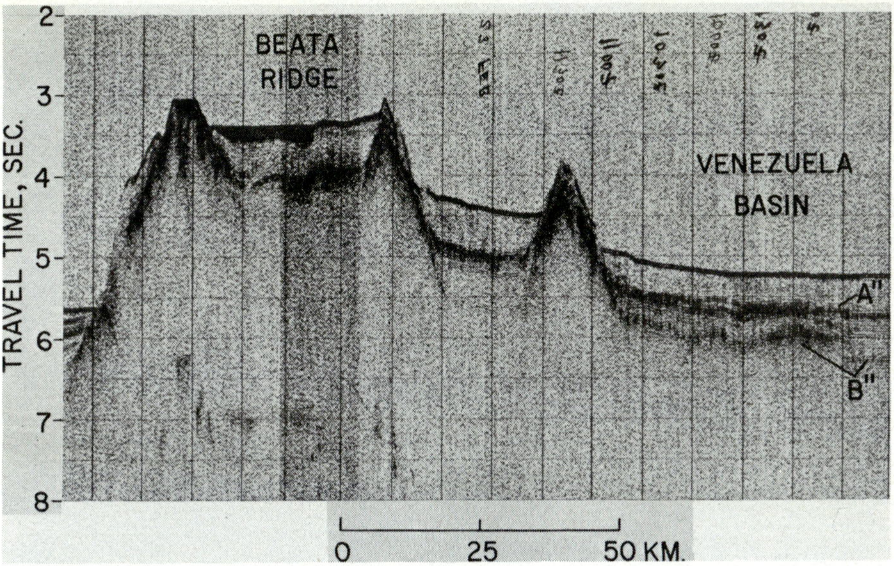

Fig.4. Northwest—southeast seismic reflection section of northern portion of Beata Ridge. Location is shown in Fig.1. Horizons A'' and B'' are labeled. Main ridge and west facing escarpment are on left. Subsidiary step-like ridges on east flank are on right. Note the thinning of sediments over the ridge crests, and the difference in thickness in the basin between the two ridges at right and the Venezuela Basin. Also note the lack of structural displacements in the sediments above the horizon B'' adjacent to the ridges, suggesting that the major structural disturbance which formed the Beata Ridge probably occurred prior to or soon after deposition of the earliest sediments overlying horizon B''.

Ridge demonstrate early Tertiary volcanism (Marlowe, 1971; Fox et al., 1971; Nagle, 1972). As on the Beata Ridge, sediment cover thins over the major structural ridges and seamounts. Bottom photographs which indicate that only coarse-grained sediments are being deposited on the crests of seamounts (Marlowe, 1971), and fine textured topographic irregularities which occur on the upper elevations of the ridge (Fig.5), suggest that the thinning of sediments at higher elevations may be the result of bottom currents. The sediment section involved in the thinning over structural highs averages at least 0.8 sec thick away from the highs, and over a substantial part of the ridge this section appears to be structurally undisturbed.

Drilling has shown that mean sedimentation rates for the Aves Ridge are 2.5—5 cm/1000 yr. for Miocene-to-recent sediments (Bader et al., 1970; Edgar et al., 1973a). Even if these unusually high rates of deposition were prevalent in the earlier part of the section, the section penetrated with seismic reflection would represent most of Cenozoic time. Thus the Aves Ridge probably has not undergone major structural dislocation since at least early Tertiary time, and from this it follows that the structural event which produced the plateau of the Aves Ridge must have occurred in pre-early Tertiary time.

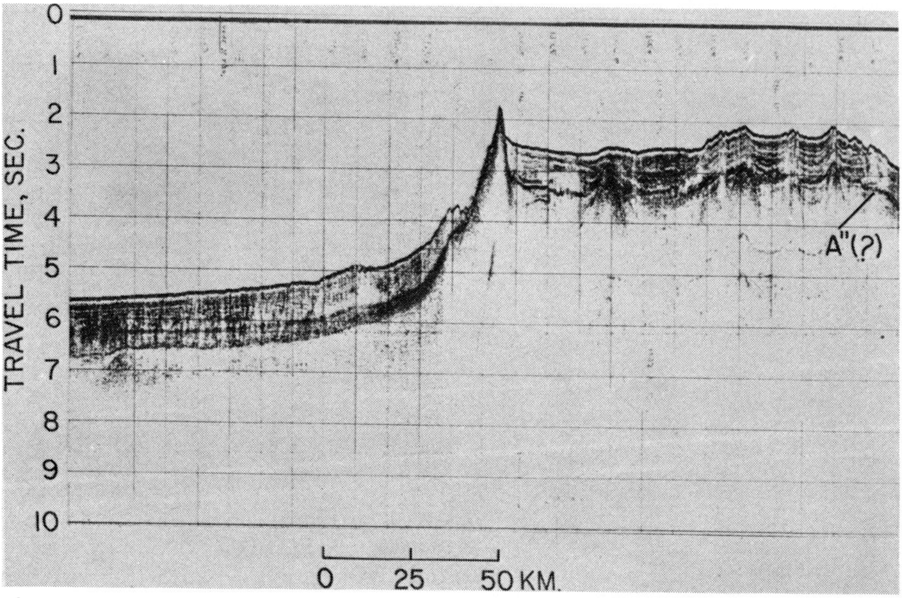

Fig.5. West-to-east seismic reflection section of Aves Ridge. Location shown in Fig.1. Note thinning of sediments over structural highs, and thickening between the highs. Also note the absence of major structural displacement in the sediment section overlying the structures. Minor structural irregularities displace the sediment section at right, hence, these could be young. A gravity slump structure occurs just west of the principal ridge. Otherwise the sediments are relatively undisturbed; they onlap, or are truncated by, the core structures. The conclusion is that, except for minor structural displacements which occurred later,the principal structures of the Aves Ridge must have been in place prior to deposition of the acoustically observed sediments.

Evidence of present-day current activity on the upper elevations of the Aves Ridge has been cited previously. Bottom photographic evidence for strong present-day current activity has also been obtained from the crest of the Beata Ridge (Fox et al., 1970). Unconformities revealed by drilling show that in fact stronger currents swept the ridge crests in late Cretaceous and early Tertiary time. Near the crest of the Beata Ridge at DSDP drill site 151 (Edgar et al., 1973a), sediments representing late Santonian-to-early Paleocene and middle Paleocene-to-middle Oligocene are missing, a time range not greatly different from that at drill site 150 in the Venezuela Basin. The occurrence of the unconformities, underlain by a "hard ground" horizon, is believed to imply "long exposure to a moderate to strong current in an oxidizing environment, possibly on a topographic high" (Edgar et al., 1973a). Similarly, a hiatus probably is present on the crest of the Aves Ridge at DSDP drill site 148. Here, volcanic sands and clay with a mixed fossil assemblage of Miocene through late Cretaceous age, lie directly below early Pliocene sediments. This presumed stratigraphic break is "marked by a brown phosphatic iron oxide that implies submarine or subaerial weathering during a period of nondeposition or erosion" (Saunders et al., 1973). Occurrence of reworked late Cretaceous nannofossils suggests a substantial time range represented by the unconformity. The seismic reflector which probably marks the unconformity is buried beneath one second or more of overlying sediments away from the structural high where the drill site was located. At DSDP drill site 30, drilled on the Aves Ridge but away from one of the structural highs, 430 m of drilling penetrated middle Miocene sediments (Bader et al., 1970).

Hiatuses or condensed sedimentary sections probably occur up and down the length of the Beata and Aves ridges, judging from the occurrence of thin sediments over the structural highs (Fig.6). The drilled sites where hiatuses were detected do not appear to be atypical of the ridges upon which they are located. It seems likely, therefore, that we are observing the geologic effects of a current of width comparable to the width of the Caribbean. As in the Venezuela Basin, water circulation across the Beata and Aves ridges must have been more intense in late Cretaceous-to-Miocene time than it is today.

CRETACEOUS-TO-MIOCENE WATER CIRCULATION

In reference to present-day water circulation, Gordon (1967) has pointed out that the Caribbean area "lacks the temperature extremes and the pronounced dominance of evaporation over precipitation for widespread thermohaline convection". Shallow sills prevent influx of cold, deep bottom water. Therefore, the major driving force for water movement in the Caribbean is the wind (ibid.). In the eastern Caribbean, the confining effect of the north and south bounding walls inhibit Ekman drift, and the resulting current flow is east to west, parallel to the prevailing wind (ibid.). Under these conditions, the frictional effects of wind stress at the surface may be propagated to greater depths than those which are possible with full development of Ekman drift. Westward current flow is therefore most intense along an axis which extends through the southern part of the eastern Caribbean, known as the Caribbean current (Wust, 1964). Water movement in this current extends to a depth of 1200—1500 m (Gordon, 1967), sufficient to affect deposition on the highest ridges and pedestals of the Aves and Beata ridges.

A westward direction of water flow at the bottom in late Cretaceous-to-Miocene time may be indicated by the position of the depositional minimum in the Venezuela Basin with respect to the underlying structure. The inferred current axis is offset to the south of the axes of the broad structural highs, though the two are parallel (Figs.2 and 3). Therefore, water flow in a westward direction is suggested, taking into consideration the dextral deflection of moving objects by the Coriolis force and assuming that in the late Cretaceous-to-Miocene the Caribbean was in the northern hemisphere. The northeast-trending wedge of sediments which cuts across the depositional minimum (Fig.3) lies within a shallow fault-bounded graben. Apparently the current was not sensitive to minor topographic irregularities,

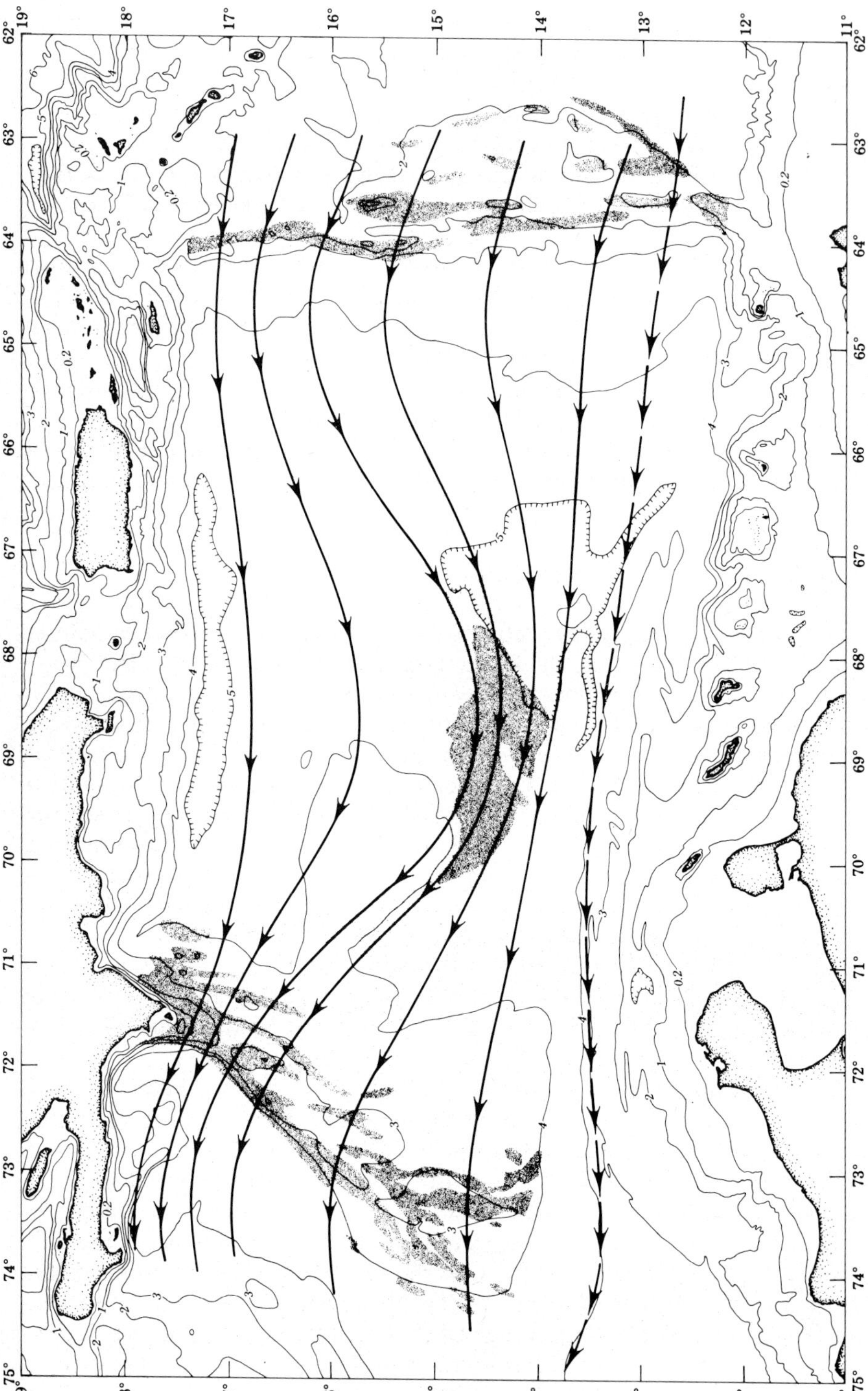

62°
63°
64°
65°
66°
67°
68°
69°
70°
71°
72°
73°
74°
75°
19°
18°
17°
16°
15°
14°
13°
12°
11°

as the axis of minimum deposition cuts across a fault trough, but it was sensitive to the location of the broad structural highs.

The extent to which the dynamics of present-day circulation can be applied to the geologic past in the eastern Caribbean is a function of how far back in time one can reasonably assume that the physiography of the eastern Caribbean, its shape, latitude, structural prominences and bounding walls, approximates that which exists at present. A reconstruction of paleolatitudes, based on inferred migration of the magnetic poles, shows the Caribbean to have been in the low latitudes of the northern hemisphere for the duration of late Cretaceous and Cenozoic time, an obvious requisite for long-continued stability of the wind-stress field (Phillips and Forsyth, 1972). The existence of the Beata and Aves ridges as structural prominences dating from late Cretaceous time has been discussed at length in the previous section of this paper.

Stratigraphic evidence from Puerto Rico, the Virgin Islands, and Hispaniola demonstrate that a shallow-water, partially emergent northern boundary for the eastern Caribbean was in existence by the middle of Cretaceous time (Donnelly, 1964; Khudoley and Meyerhoff, 1971) and that it probably persisted through late Cretaceous and Cenozoic time. While sills breaching the Greater Antilles Island Arc may have been in existence, it is probable that water movement through the arc has been restricted since mid-Cretaceous time, and possibly at times it was limited severely or cut off altogether.

The north—south width of the Venezuela Basin in Campanian-to-Miocene time probably varied as the plates of the region moved with respect to one another. A reconstruction of the movements of the South American plate relative to the North American plate, based on paleomagnetic evidence of sea-floor accretion (Ladd, 1974), would have the latitudinal extent of the Caribbean increasing during late Cretaceous—Eocene time and decreasing during Eocene—Miocene time.

Both the northern and southern margins of the Venezuela Basin exhibit the probable effects of convergent crustal blocks or lithospheric plates (Garrison et al., 1972; Matthews and Holcombe, 1974; Case, 1975; Silver et

---

Fig.6. Sketch of streamline deflection which might occur in a westward flowing current moving across the Aves Ridge, Venezuela Basin, and Beata Ridge. Sketch is qualitative, based on model for streamline deflection over a topographic barrier developed by Neumann (1960), which takes into account Coriolis and frictional effects. In Neumann's model the maximum deflection occurs downstream of the topographic barrier. Streamlines have been drawn in to coincide with the axis of minimum deposition in the Venezuela Basin. Current flow on the floor of the Venezuela Basin has not been completely independent of bottom topography, as illustrated in Fig.2. Areas of appreciable sediment thinning (<0.5 sec two-way travel time) are shown by dark shading. Areal extent of Venezuela Basin "depositional scar" included by light shading. Because of the boundary effects of the Greater Antilles and the South American Continental Margin, maximum deflection would occur in the center of the Venezuela Basin, thereby concentrating flow in the southern part of the basin. Axis of the present Caribbean current (Gordon, 1967), probably associated with a less intense current regime, is shown by dashed flow line. Bathymetry is in uncorrected meters × 1000.

al., 1975). Thus, the Venezuela Basin was probably of greater width during most of the time interval in question than at present; however, it is unlikely that major changes in latitude and the geometry of the bounding walls and ridges of the eastern Caribbean which would drastically alter water circulation have occurred.

A greater vertical column of water transported by wind stress, or shallower Aves and Beata ridges, in pre-Miocene time, would probably result in deflection of streamlines due to the topographic effect of the Aves and Beata ridges. Such deflection, ascribed to Coriolis and frictional effects (Neumann, 1960) is believed to explain the southward deflection of the Gulf Stream over the Southeast Newfoundland Ridge. In the northern hemisphere, deflection would be to the right over a topographic barrier and to the left beyond it. Double deflection over the Aves and Beata ridges is one mechanism to explain the arcuate shape of the principal axis of sediment thinning in the Venezuela Basin, as illustrated in Fig.6. Such a flow pattern also offers an explanation for the gradual fade-out of depositional effects at the ends of the depositional axis in the Venezuela Basin, far from the ridges and in the absence of notable deepening.

An alternate explanation for the arcuate shape of the Venezuela Basin unconformity belt would be a gyre in the deep circulation of the basin which might be induced by wind or thermohaline effects or both; however, we consider this unlikely.

The current was apparently able to maintain itself through times when the Aves and Beata ridges were shallower with respect to sea level (shallower ridges could also accentuate streamline deflection). Several samples of middle Eocene neritic carbonates were obtained from cores adjacent to a peak on the Beata Ridge which has a minimum depth of about 1000 m, suggesting that on the order of this amount of subsidence has subsequently taken place (Fox et al., 1970). On the Aves Ridge, whose peaks rise to shallower depths than those of the Beata Ridge, neritic carbonates of late Eocene, Oligocene, and early to middle Miocene were dredged, suggesting a shallower ridge crest from Eocene through Miocene times (Fox et al., 1971; Bock, 1972). The depths of pedestals and ridges adjacent to the dredge hauls on the Aves Ridge requires a minimum subsidence of 200—600 m. Seismic reflection sections of the ridge strongly suggest that only the highest ridges and seamounts have been emergent or near sea level since early Tertiary time (see Fig.5). Thus it is unlikely that post-Miocene subsidence generally has been much greater than 600 m. Fig.7, a sill-depth profile along the length of the Aves Ridge crest, illustrates the effect that a shallower ridge crest would have on the water cross-section.

ALTERATION OF THE CURRENT REGIME IN THE MIOCENE

The current regime which influenced depositional patterns in the Venezuela Basin for 50 m.y. was altered or seriously weakened in early-to-middle Miocene time. Miocene-to-recent sediments apparently have accumulated on

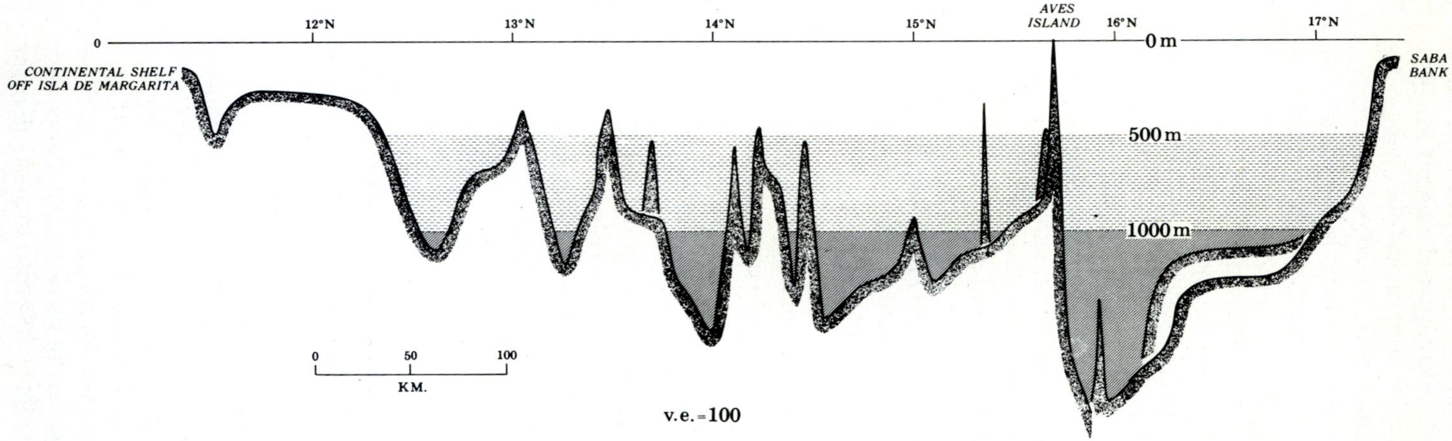

Fig.7. Profile of crest-line depth of Aves Ridge, normalized to a N—S plane. Several peaks from adjacent crests have been projected onto the plane. Shaded areas show the 500-to-1000 m and greater-than-1000 m depth intervals, thereby illustrating the change in the water cross-section which would occur as a consequence of reconstructing the Aves Ridge to a reasonable pre-subsidence level. Apparently a 500 m shallowing would still allow relatively free circulation, whereas a 1000 m shallowing would have a substantial restricting effect. Depths are in uncorrected meters (1 sec. two-way travel time = 750 m).

previously current-swept crests of the Aves and Beata ridges, and in the Venezuela Basin at drill site 150, sediments accumulated where formerly none were deposited. Outside the axis of minimum deposition, a decrease in the percentage of carbonates in the sediments occurred, and the last occurrence of Radiolaria in Caribbean marine sediments is in middle Miocene sediments (Saunder et al., 1973). What changes occurred which would account for this change in the current regime?

The late Cretaceous—early Tertiary paleocirculation in the eastern Caribbean may have been part of a more intense westward circulation through a Caribbean open to the Pacific. The Tethyan seaway, also open during pre-Miocene time, may have contributed to the intensity of the circulation. The constricting effect of the Caribbean, situated between two open oceans, no doubt contributed to the intensity of water circulation. Water flow through the Caribbean during pre-Miocene time could be somewhat analogous to the present transport of water through the Drake Passage. West-to-east water transport through the Drake Passage, driven by circum-global winds, is intense and affects the entire water column, from surface to bottom, in water depths of 3000—4000 m (Reid and Nowlin, 1971). It is probable that the constricting effect of the Drake Passage further intensifies circulation relative to that of the circum-Antarctic current in the open ocean (Gordon, 1966; Hollister and Heezen, 1967).

Another factor which could have affected the sedimentation regime was the subsidence of the Beata and Aves ridges, which occurred after Eocene time on the former and in about Miocene time on the latter.

In a rotating-tank experiment, Luyendyk et al. (1972) attempted to simulate open-Caribbean, open-Tethys water circulation. In their model a strong circum-global westward circulation was induced through the Tethyan—Caribbean region, which they called the Tethys Current (Fig.8). In spite of limitations, their model might be applicable to Caribbean paleocirculation to the extent that water movement and the resulting geologic effects were controlled by wind-driven circulation.

A marine Caribbean—Pacific connection through Panama was in existence through late Cretaceous and most of Cenozoic time. The basement rock of Panama, Costa Rica, and northwest Colombia is an oceanic basaltic assemblage containing pillow lavas and overlain by abyssal marine sedimentary rocks (Case, 1974). By Eocene time and very probably before, an island arc was present through the Panama region (Malfait and Dinkelman, 1972). The period between late Eocene and the end of Oligocene time was one of relative submergence and volcanic quiescence (Lloyd, 1963). Early Miocene was a time of emergence throughout southern Costa Rica and western Panama (ibid.). By middle Miocene time, land, probably attached to nuclear Central America, was present in western Panama (Whitmore and Stewart, 1965), but deep marine sediments were being deposited in eastern Panama (Case, 1974). In the late Miocene, uplift occurred in eastern Panama (Case, 1974), and an interchange of vertebrate fauna was taking place by late Pliocene (Stirton, 1950). Marine biostratigraphic evidence likewise suggests that the

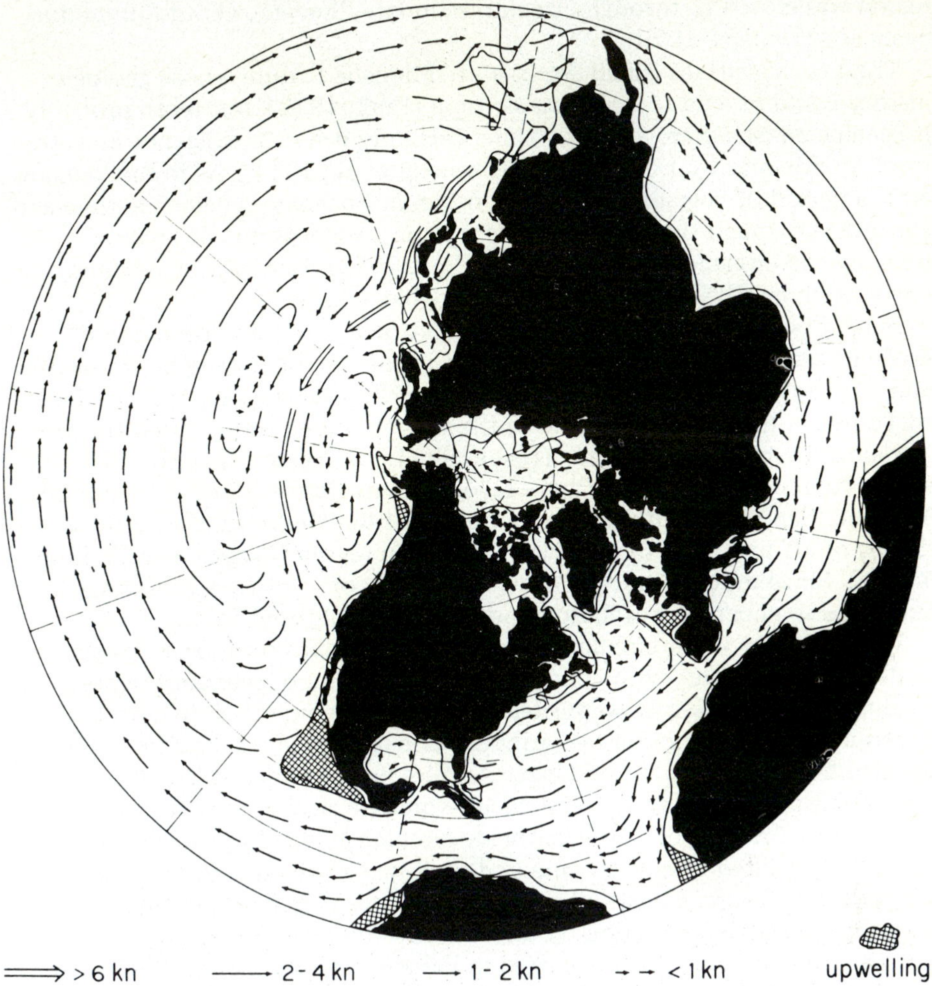

Fig.8. Reproduction of a vector presentation of the results of the rotating tank experiment of Luyendyk et al. (1972). Scale of current speeds is shown. Their model is based on the reconstruction of Atlantic-bordering continents of Phillips and Forsyth (1972). The principal features of this reconstruction are: (1) that the Tethys region was in a lower latitude (15—30° N), whereas the Caribbean region remains in a lower latitude similar to that of today (10—20° N); and (2) that open seaways extended through the Tethys and Caribbean regions. They assume that this model is applicable for the late Cretaceous and the Tertiary through Oligocene—Miocene: however, it does not allow for exchange of water between the hemispheres, and it does not consider geologic alterations which have changed the shape and position of continental block boundaries, such as the rotations and relative movements of small plates, or growth of a continental margin through uplift and orogenesis of a continent-bordering wedge of sediments. In spite of these and other limitations, they were able to reproduce the gross features of present oceanic circulation by configuring the scale-model continents as they are now positioned on the earth. Reprinted by permission of the Geological Society of America.

region was closed to through circulation in late Pliocene, about 3.5 million years ago (Kaneps, 1970).

The Lesser Antilles Island Arc is a pre-Miocene feature whose geologic history could extend to late Jurassic time (Fink, 1972), but more probably, it became an island arc at a later time; certainly it was in existence as a structural prominence by Eocene time (Donnelly et al., 1971). As in the Panama area, a period of volcanic quiescence began in Eocene or Oligocene time and continued through the early Miocene; this interval was one of erosion, possibly accompanied by subsidence (Fink, 1972). Renewal of volcanism occurred in middle or late Miocene time (ibid.).

Since Eocene time the history of the Tethyan region has been one of gradual restriction of the seaway, caught in the vise of converging Eurasian and African plates (Phillips and Forsyth, 1972; Dewey et al., 1973). Closure of Tethys in the Middle East and in the Gibraltar area had isolated the Mediterranean by late Miocene time (Hsu et al., 1973; Nesteroff, 1973). The benthonic foraminiferal fauna bear the record of this closure. Prior to the middle Miocene, Cenozoic benthonic foraminifera of the Mediterranean show a high degree of affinity with those of the Caribbean region (Berggren and Phillips, 1971*). Since middle Miocene time, the faunas of the two regions have been subjected to geographic isolation and separate evolution (ibid.).

Weakening of water circulation through the eastern Caribbean in early-to-middle Miocene time, with concomitant resumption of sediment deposition in areas which previously were sites of nondeposition, seems to correlate generally with constriction of Caribbean and Tethys and specifically with constriction or closure in the Lesser Antilles and in Panama. The decline of circulation appears to chronologically predate final closure in Panama. However, restriction of Caribbean-wide circulation would probably be more reasonably related to constriction and shallowing, which could occur long before final closure. Constriction in Panama probably reached the point of impeding through circulation by middle Miocene time. Subsidence and volcanic quiescence, which occurred in Panama and the Lesser Antilles in late Eocene-to-early Miocene, correlates with the time during which the unconformity reached its maximum extent (Fig.2). Renewal of uplift and volcanic activity in Panama and the Lesser Antilles in middle Miocene time correlates well with the upper limit of the unconformity.

We have invoked the possibility of strong east-to-west wind-driven circulation through the Caribbean/Tethys belt as an explanation for late Cretaceous-to-Miocene nondeposition on the floor of the eastern Caribbean. Away from the confining effect of the Caribbean "bottleneck" to circulation, in the open Atlantic and Pacific, where the Ekman effect would have been more fully developed, one would not expect to see the track of wind-driven circulation recorded in bottom sediments. One possible exception is the occur-

---

*Those authors suggest that the most probable means of transportation of the benthonic Foraminifera across the Atlantic was by ocean currents. A strong Tethys current could have provided the means for this faunal interchange.

rence of an unconformity of similar time range which has been drilled in deep water in the eastern Atlantic, west of Gibraltar. (Drill site 136, Pimm and Hayes, 1972). Here deep-sea pelagic sediments lie above and below a late Cretaceous-to-Miocene unconformity.

Prior to onset of strong currents in the Caribbean region, quiet water conditions alternating with very weak circulation were prevalent, not only in the Caribbean but in the Atlantic as well. Saunders et al. (1973) have concluded that initiation of strong circulation in the Caribbean and equatorial Atlantic regions resulted when drift of the South American plate apart from the African plate opened the South Atlantic to the North Atlantic west of the African bight. Perhaps it was opening of the equatorial Atlantic, combined with an already open Tethys/Caribbean, which provided the conditions for a strong circulation through the Caribbean and possibly throughout much of the Atlantic-Tethys region, and this condition was maintained until the termination of east-to-west global circulation through the Tethys/ Caribbean region.

ACKNOWLEDGEMENTS

This paper is an outgrowth of earlier collaborative study with N.T. Edgar of Scripps Institution of Oceanography, J.I. Ewing of Lamont-Doherty Geological Observatory of Columbia University, and W.R. Johnson of the U.S. Naval Oceanographic Office, which resulted in the synthesis of acoustic reflection data with D.S.D.P. drillhole stratigraphy (Edgar et al., 1973b). Discussions of the subject matter with W.F. Ruddiman, J.E. Matthews, E.C. Escowitz, L. Kovacs, L.K. Fink and P.J. Fox have been most helpful and stimulating. Delineation of sediment thickness patterns in the eastern Caribbean was made possible through the use of seismic reflection data kindly made available by J.I. Ewing and W.J. Ludwig of Lamont-Doherty Geological Observatory, N.T. Edgar of Scripps Institution of Oceanography, D.A. Fahlquist and W.R. Bryant of Texas A & M University, J.E. Case of the U.S. Geological Survey, and L.K. Fink of the University of Maine. The seismic reflection sections shown in Figs.4 and 5 were made by U.S.N.S. "Wilkes" of the U.S. Naval Oceanographic Office. We thank A.R. Gordon, W.F. Ruddiman, J. Helwig, N.T. Edgar, and P.J. Fox for reviewing the manuscript. P. Michalco drew up the illustrations. This work is a result of the Caribbean Study Program being conducted by the U.S. Naval Oceanographic Office.

REFERENCES

Bader, R.G. et al., 1970. Site Reports, Part I. In: Initial Reports of the Deep-Sea Drilling Project, IV. U.S. Government Printing Office, Washington, D.C., pp.2—366.

Bender, M.L., Ku, T.L. and Broecker, W.S., 1966. Manganese nodules: their evolution. Science, 151: 325—328.

Berggren, W.A. and Phillips, J.D., 1971. Influence of continental drift on the distribution of the Tertiary benthonic Foraminifera in the Caribbean and Mediterranean Regions.

In: Symposium on the Geology of Libya. Faculty of Science, Univ. of Libya, Benghazi, pp.263—299.

Bock, W.D., 1972. The use of Foraminifera as indicators of subsidence in the Caribbean. In: Trans. 6th Caribbean Geol. Conf. Queens College Press, New York, N.Y., pp.439—440.

Case, J.E., 1974. Oceanic crust forms basement of eastern Panama. Geol. Soc. Am. Bull., 85: 645—652.

Case, J.E., 1975. Geophysical studies in the Caribbean Sea. In: A.E.M. Nairn and F.G. Stehli (Editors), Ocean Basins and Margins, 3. The Caribbean and Gulf of Mexico. Plenum, New York, N.Y., pp. 107—180.

Case, J.E. and Holcombe, T.L., 1975. Preliminary geologic-tectonic map of the Caribbean region. U.S. Geol. Surv. Open-File Map, 75—146.

Dewey, J.F., Pitman III, W.C., Ryan, W.B.F. and Bonnin, J., 1973. Plate tectonics and the evolution of the Alpine System. Geol. Soc. Am. Bull., 84: 3137—3180.

Donnelly, T.W., 1964. Evolution of eastern Greater Antillean Island Arc. Bull. Am. Assoc. Pet. Geol., 48: 680—696.

Donnelly, T.W., Rogers, J.J.W., Pushkar, P. and Armstrong, R.L., 1971. Chemical evolution of the igneous Rocks of the eastern West Indies: an investigation of thorium, uranium, and potassium distributions, and lead and strontium isotopic ratios. Geol. Soc. Am. Mem., 130: 181—224.

Edgar, N.T., Ewing, J.I. and Hennion, J., 1971. Seismic refraction and reflection in the Caribbean Sea. Bull. Am. Assoc. Pet. Geol., 55: 833—870.

Edgar, N.T., Saunders, J.B. et al., 1973a. Site Reports, Part I. In: Initial Reports of the Deep-Sea Drilling Project, XV. U.S. Government Printing Office, Washington, D.C., pp.3—472.

Edgar, N.T., Holcombe, T.L., Ewing, J.I. and Johnson, W.R., 1973b. Sedimentary hiatuses in the Venezuelan Basin. In: Initial Reports of the Deep-Sea Drilling Project, XV. U.S. Government Printing Office, Washington, D.C., pp.1051—1062.

Ewing, J.I., Talwani, M., Ewing, M. and Edgar, N.T., 1967. Sediments of the Caribbean. In: Studies in Tropical Oceanography 5. Univ. of Miami, Coral Gables, Fla., pp.88—102.

Ewing, J.I., Talwani, M. and Ewing, M., 1968. Sediment distribution in the Caribbean Sea. In: Trans. 4th Caribbean Geol. Conf. Queens College Press, New York, N.Y., pp. 317—323.

Fink Jr., L.K., 1972. Bathymetric and geologic studies of the Guadeloupe region, Lesser Antilles Island Arc. Mar. Geol., 12: 267—288.

Fox, P.J. and Heezen, B.C., 1975. Geology of the Caribbean Crust. In: A.E.M. Nairn and F.G. Stehli (Editors), Ocean Basins and Margins, 3. The Caribbean and Gulf of Mexico. Plenum, New York, N.Y., pp. 421—466.

Fox, P.J., Ruddiman, W.F., Ryan, W.B.F. and Heezen, B.C., 1970. The geology of the Caribbean crust, I: Beata Ridge. Tectonophysics, 10: 495—513.

Fox, P.J., Schreiber, E. and Heezen, B.C., 1971. The geology of the Caribbean crust: Tertiary sediments, granitic and basic rocks from the Aves Ridge. Tectonophysics, 12: 89—109.

Garrison, L.E. et al., 1972. USGS-IDOE Leg 3. Geotimes, 17(3): 14—15.

Gordon, A.L., 1966. Potential temperature, oxygen, and circulation of bottom water in the Southern Ocean. Deep-Sea Res., 13: 1125—1138.

Gordon, A.L., 1967. Circulation of the Caribbean Sea. J. Geophys. Res., 72: 6207—6223.

Heezen, B.C. and Hollister, C.D., 1971. The Face of the Deep. Oxford University Press, New York, N.Y., 650 pp.

Hollister, C.D. and Heezen, B.C., 1967. The Floor of the Bellingshausen Sea. Johns Hopkins Oceanogr. Stud., 3: 177—189.

Hollister, C.D. and Heezen, B.C., 1973. Geologic effects of ocean bottom currents in the western North Atlantic. In: A.L. Gordon (Editor), Studies in Physical Oceanography, 2. Gordon and Breach, New York, N.Y., pp.37—66.

Holcombe, T.L. and Matthews, J.E., 1973. Structural fabric of the Venezuelan Basin,

Caribbean Sea. Geol. Soc. Am., Abstr. Progr., pp.671—672.
Holcombe, T.L., Edgar, N.T. and Matthews, J.E., 1973. Sediment distribution and structure in the Venezuelan Basin. Annu. Meet. Southwest Sec., Am. Assoc. Pet. Geol., Abstr. Progr.
Horn, D.R., Horn, B.M. and Delach, M.N., 1972. Distribution of ferromanganese deposits in the World Ocean. In: D.R. Horn (Editor), Papers from a Conference on Ferromanganese Deposits on the Ocean Floor. International Decade of Ocean Exploration. National Science Foundation, Washington, D.C., pp.9—17.
Hsu, K.J., Cita, M.B. and Ryan, W.B.F., 1973. The origin of the Mediterranean evaporites, IV. In: Initial Reports of the Deep-Sea Drilling Project, XIII (part 2). U.S. Government Printing Office, Washington, D.C., pp.1203—1231.
Kaneps, A.G., 1970. Late Neogene Biostratigraphy, Biogeography, and Depositional History. Thesis, Columbia Univ., New York, N.Y., 185 pp., unpublished.
Khudoley, K.M. and Meyerhoff, A.A., 1971. Paleogeography and geological history of the Greater Antilles. Geol. Soc. Am. Mem., 129: 199 pp.
Krishnaswami, S., Somayajulu, B.L.K. and Moore, W.S., 1972. Dating of manganese nodules using beryllium-10. In: D.R. Horn (Editor), Papers from a Conference on Ferromanganese Deposits on the Ocean Floor. International Decade of Ocean Exploration. National Science Foundation, Washington, D.C., pp.117—122.
Ladd, J., 1974. Relative motion between North and South America and Caribbean tectonics. In: VIIème Géol. Conf. Caribbéan, Française Antilles. Recueil des Résumés et Communications, 37 (abstract).
Lloyd, J.J., 1963. Tectonic history of the south central-American orogen. Am. Assoc. Pet. Geol. Mem., 2: 88—100.
Luyendyk, B.P., Forsyth, D. and Phillips, J.D., 1972. Experimental approach to the paleocirculation of the oceanic surface waters. Geol. Soc. Am. Bull., 83: 2649—2664.
Malfait, B.T. and Dinkelman, M.G., 1972. Circum-Caribbean tectonic and igneous activity and the evolution of the Caribbean Plate. Geol. Soc. Am. Bull., 83: 251—272.
Marlowe, J.I., 1971. Geological reconnaissance of parts of the Aves Ridge. In: Trans. 5th Caribbean Geol. Conf., Queens College Press, New York, N.Y., pp.61—64.
Matthews, J.E. and Holcombe, T.L., 1974. Possible Caribbean underthrusting of the Greater Antilles along the Muertos Trough. In: VIIème Géol. Conf. Caribbéan, Française Antilles. Recueil des Résumés et Communications, 44 (abstract).
Moore, W.A., 1973. Accumulation rates of manganese crusts on rocks exposed on the sea floor. In: Phase I Report, Inter-University Program of Research on Ferromanganese Deposits of the Sea Floor. Seabed Assessment Programs, International Decade of Ocean Exploration. National Science Foundation, Washington, D.C., pp.93—97.
Nagle, F., 1972. Rocks from seamounts and escarpments on the Aves Ridge. In: Trans. 6th Caribbean Geol. Conf., Queens College Press, New York, N.Y., pp.409—413.
Nesteroff, W.D., 1973. The sedimentary history of the Mediterranean area during the Neogene, Part IV. In: Initial Reports of the Deep-Sea Drilling Project, XIII (part 2). U.S. Government Printing Office, Washington, D.C., pp.1257—1261.
Neumann, G., 1960. On the effect of bottom topography on ocean currents. Dtsch. Hydrogr. Z., 13: 132—141.
Phillips, J.D. and Forsyth, D., 1972. Plate tectonics, paleomagnetism, and the opening of the Atlantic. Geol. Soc. Am. Bull., 83: 1579—1600.
Pimm, A.C. and Hayes, D.E., 1972. General Synthesis, Part III. In: Initial Reports of the Deep-Sea Drilling Project, XIV. U.S. Government Printing Office, Washington, D.C., pp.955—975.
Reid, V.L. and Nowlin, W.D., 1971. Transport of water through the Drake Passage. Deep-Sea Res., 18: 51—64.
Roemer, L.B., Bryant, W.R. and Fahlquist, D.A., 1973. Geology and geophysics of the Beata Ridge—Caribbean. Texas A & M Univ. Tech. Rep. 73—14—T: 92 pp.
Saunders, J.B., Edgar, N.T., Donnelly, T.W. and Hay, W.W., 1973. Cruise synthesis, Part III. In: Initial Reports of the Deep-Sea Drilling Project, XV. U.S. Government Printing

Office, Washington, D.C., pp.1077—1111.
Silver, E.A. et al., 1972. Acoustic Reflection Profiles, Venezuelan Continental Borderland. International Decade of Ocean Exploration and U.S. Geol. Surv. Rep. No. USGS-GD-72-005.
Silver, E.A., Case, J.E. and MacGillavry, H.J., 1975. Geophysical study of the Venezuelan Borderland. Geol. Soc. Am. Bull., 86: 213—226.
Stirton, R.A., 1950. Late Cenozoic avenues of dispersal for terrestrial animals between North America and South America. Geol. Soc. Am. Bull., 61: 1541—1542 (abstract).
Whitmore Jr., F.C. and Stewart, R.H., 1965. Miocene mammals and central American seaways. Science, 148: 180—185.
Wust, G., 1964. Stratification and Circulation in the Antillean—Caribbean Basins. Columbia University Press, New York, N.Y., 201 pp.

Marine Geology, 23 (1977) 57—75

ABYSSAL BEDFORMS EXPLORED WITH A DEEPLY TOWED INSTRUMENT PACKAGE*

PETER LONSDALE and F.N. SPIESS

University of California, San Diego, Calif. (U.S.A.)
Marine Physical Laboratory of the Scripps Institution of Oceanography, La Jolla, Calif. (U.S.A.)

(Received April 28, 1976)

ABSTRACT

Lonsdale, P. and Spiess, F.N., 1977. Abyssal bedforms explored with a deeply towed instrument package. Mar. Geol., 23: 57—75.

Several types of abyssal bedforms have been discovered during surveys with a deeply towed instrument package at water depths of 1.5—6 km in the Pacific and Atlantic Oceans. Cores and current-meter records obtained at the same sites provide data for interpreting their dynamics. Wave and current ripples are best portrayed in bottom photographs, but medium-scale bedforms, including sand waves, mud waves and erosional furrows, are described by interpreting high-resolution side-looking sonar records. The largest examples affect surface-ship echograms, though their shape and structure can seldom be resolved without near-bottom observations. Wave ripples are common on the slopes of seamounts and ridges, while current ripples and sand waves occur beneath some fast thermohaline currents whose beds are shallower than the foraminiferal compensation depth. Depositional and erosional bedforms in cohesive sediment have been found beneath the deepest thermohaline currents; they may be restricted to areas where the flow is unusually steady in direction.

INTRODUCTION

The discovery of sedimentary bedforms at abyssal depths that geological and oceanographic dogma held to be free of fast, sediment-moving currents was a major achievement of early deep-sea photography (e.g., Menard, 1952). Many thousands of photographs taken during the past three decades show current ripples and oriented scour around obstacles; numerous investigators have used them to interpret the texture and structures of sediment cores (e.g., Heezen and Hollister, 1964; Hollister and Heezen, 1971; Zimmerman, 1972) and the routes and relative speed of the abyssal circulation (e.g., Hollister and Elder, 1969; Goodell et al., 1971; Heezen and Hollister, 1971). The validity of this latter technique has often been demonstrated by subsequent direct measurement of currents first inferred from photographic evi-

*Contribution of the Scripps Institution of Oceanography, new series.

dence, for example on the North American continental rise and the Blake-Bahama Outer Ridge. "Snapshot hydrography" relies on an assumed understanding of the relationship of such bedforms as current ripples and scour crescents to the flow regime, an understanding derived from a century of study of their shallow marine and fluviatile analogues. However, some of the bedforms discovered by deep-sea photography (e.g., large sharp-crested longitudinal ripples) seem rather different from any well-studied shallow-water bedforms, and their significance is obscure. The large-scale corrugations, with wavelengths of 0.5—5 km, that cover wide areas near some continental margins (e.g., Fox et al., 1968; Rona, 1969; Egloff, 1972), and have been mapped with echosounders rather than with cameras, also have an uncertain significance; 3.5-kHz records demonstrate that some of these features are "giant mud waves", with large-scale cross-bedding indicative of prograding bedforms (e.g., Ewing et al., 1971), although some may be slump structures unaffected by bottom currents (Lonsdale, 1975a). Ignorance of their dynamics limits the oceanographic inferences that can be derived from distribution patterns.

Our approach to the study of abyssal bedforms begins with mapping small patches of them in great detail, measuring near-bottom currents over them, and coring to determine the composition of their sediments. Using these observations, we try to establish whether the bedforms are in equilibrium with the observed flow, are created by episodic flows with a long recurrence interval, or are relics from some previous flow regime. Our long-term goal is to determine the equilibrium flow conditions for each type of abyssal bedform, so that we can predict their distribution beyond the tiny patches we survey, or, conversely, use observations of bedforms or their marks in the geologic record to make more precise inferences about the flows prevailing during their formation. The limited aim of this brief review is to describe some of the abyssal bedforms we have explored with our novel instrumentation system, concentrating on those that cannot be studied adequately with conventional deep-sea photographic systems or surface-ship echosounders.

METHODS OF STUDY

Our principal mapping tool has been the deeply towed instrument package of the Marine Physical Laboratory of the Scripps Institution of Oceanography (Spiess and Mudie, 1970; Spiess and Tyce, 1973), which is towed 10—100 m above the sea floor (at depths as great as 7500 m) at speeds of 2—4 km/hr and is navigated within an acoustic transponder net. Of its several sensor systems, those most useful for studies of bedforms include:

(1) A set of vertically mounted cameras, arranged in a variety of configurations that generally allows us to take stereo-pairs and a continuous 10-m-wide swath of overlapping photographs. The cameras are fired by remote command from the towing ship, and contain 1200 frames of 35-mm film (usually black-and-white, though we have had some recent success with color film). A snapshot TV system provides an immediate image of the sea floor when each frame is shot.

(2) A pair of side-looking sonars, operating at 110 kHz, which record the acoustic backscattering properties of the sea floor out to ranges of about 500 m on either side of the vehicle (at typical operating altitudes).

(3) A 40-kHz echosounder and a 4-kHz seismic profiler, which provide high-resolution data on bathymetry and shallow subbottom reflecting horizons.

Bottom currents within our survey areas are measured with arrays of free-vehicle current meters, designed by the Marine Life Research Group of the Scripps Institution of Oceanography (Schick et al., 1968). They are usually deployed attached to our navigational transponders as part of a multipurpose free-vehicle package.

We obtain sediment samples with conventional corers. Depending on how precisely positioned we require the sample to be, we use free-fall or piston corers from a transponder-positioned ship (with a resulting uncertainty in relative position of 100—200 m) or navigate piston and gravity corers all the way to the sea floor with a relay transponder-navigation system, which allows a positioning error of less than 20 m (Boegeman et al., 1972).

Since 1967, deep-tow surveys have been conducted at many scattered but carefully selected sites throughout the Pacific and North Atlantic Oceans (Fig.1). At those sites where abyssal sedimentation is our principal interest (Table I), the typical pattern of observations has been to conduct a surface-ship site survey; deploy an array of transponders and current meters; lower the deep-tow instrument and tow it in looping patterns within the array (interrupting the lowering as necessary to recharge cameras with film); then core targets selected on the basis of the deep-tow observations; and finally retrieve the transponders and current meters. The main disadvantages of this routine are that the current-meter records are shorter than desirable (Table I), and often not from the ideal spot, because the meters are set before the deep tow has established the distribution of bedforms, which are usually undetectable on surface-ship records.

CLASSIFICATION AND DEFINITIONS

Several classifications of bedforms have been proposed (e.g., Potter and Pettijohn, 1963; Allen, 1968; Reineck and Singh, 1973), using as criteria supposed origin, morphology and orientation relative to the flow direction, size, and composition. Some of these characteristics are so poorly determined for abyssal bedforms that we have not fitted our descriptions into a structure erected around shallow-water phenomena. Instead of a classification, we use a pragmatic division into:

(1) *Ripples*: small-scale regular alternations of depositional ridges a few centimeters apart. They may be symmetric or asymmetric, transverse or longitudinal to the flow, and are formed by currents flowing over unconsolidated sand, silt or mud. Best portrayed on bottom photographs, they are seldom resolved by side-looking sonars.

(2) *Sand Waves* or dunes: large-scale counterparts of sand ripples, with typical lateral dimensions of 10—100 m. All known abyssal examples are

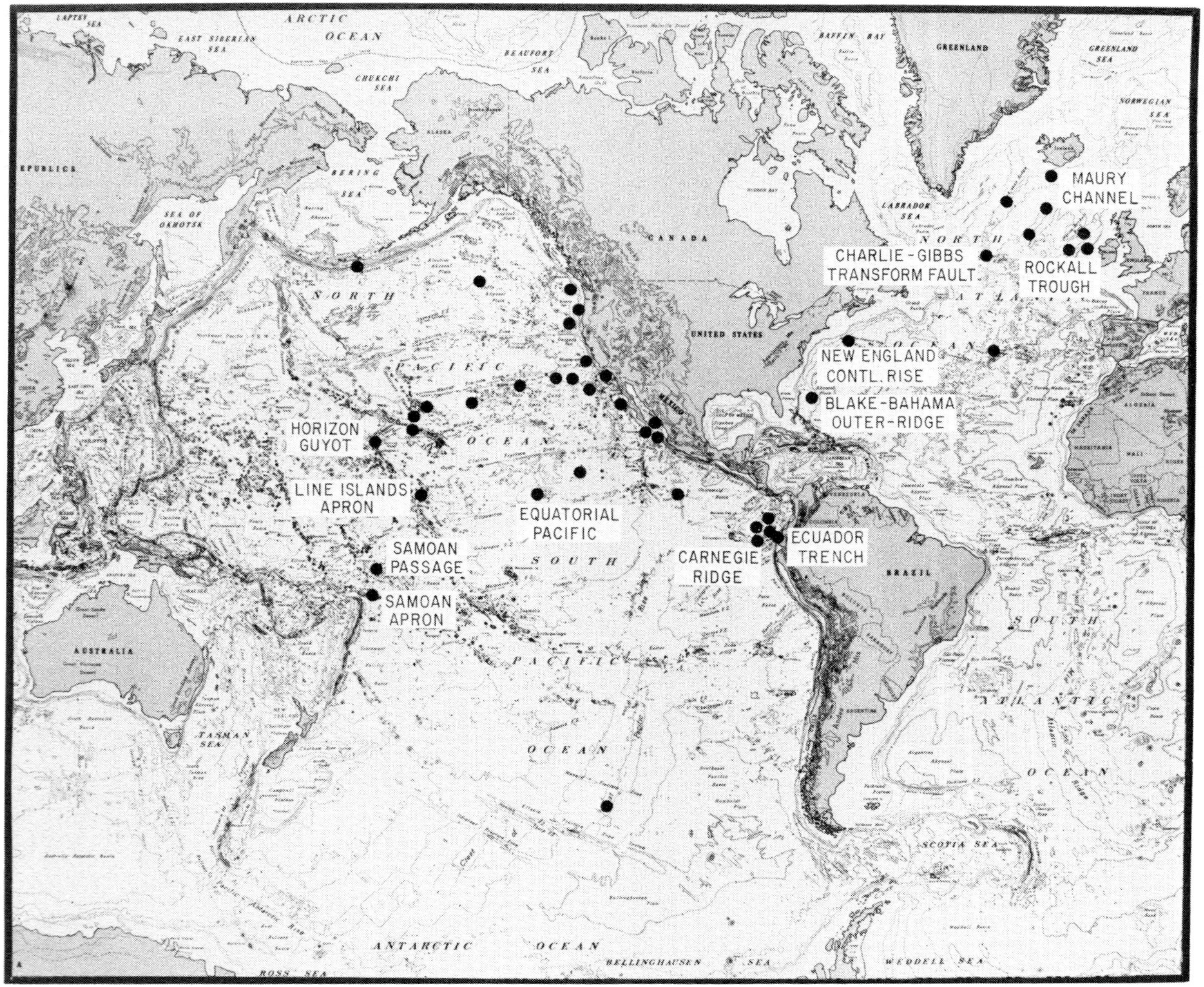

Fig.1. Location of deep-tow survey areas. The principal surveys where we have studied abyssal sedimentation processes and found bed-forms are named.

TABLE I

Deep tow studies of abyssal sedimentation

Year of survey	Area	No. of current meter records	Duration of records (days)	Highest average velocity (cm/sec)	Highest max. speed*[1] (cm/sec)	Bedforms and obstacle marks	References*[3]
1969, 1970	Equatorial Pacific, 134°W.	3	5	5.4	10	none	1, 2
1970	Horizon Guyot	3	2–5	2.5	17	sand ripples, dunes	3
1972	South Carnegie Ridge	3	3–5	3.0	15	sand ripples, dunes	4
1972, 1974	North Carnegie Ridge	3	2–16	1.4	19	sand ripples, dunes; streamers, patterned nodules	5
1972	Samoan Apron	1	30	0.8	10	none	6
1972	Samoan Passage (4 sites)	24	3–10	17.0	26	furrows, patterned nodules; large moats	7
1973	Line Islands Apron	2	2–9	2.4	10	sand ripples; large moats	8
1973	Blake-Bahama Outer Ridge	6	4–6	5.0	11	furrows, mud "ripples"	9, 10
1973	New England Contl. Rise	3	4	8.4	20	mud waves; large moat, crescents	
1973	New England Seamount	1	2	4.3	20	ripples, crescents, shadows	11
1974	East Carnegie Ridge	3	4	5.7	28	large scoured depressions	12
1974	Ecuador Trench	4	5–17	34.4	40	sand ripples	12, 13
1975	Charlie-Gibbs Transform Fault	6	2–3	7.4	21	mud ripples; crescents	16
1975	Maury Channel (3 sites)	3	1	14.6	20	crescents	15, 16
1975	Hatton Drift (2 sites)	2	7	13.8	21	sand ripples, dunes; crescents	15
1975	Rockall Trough (4 sites)	2*[2]	4–5	3.7	14	sand ripples, dunes, crescents, shadows	17

*[1] Maximum speeds averaged over 15 minutes.
*[2] Towing in Rockall Trough included a long transect; current meters were not at same site as bedforms.
*[3] References: *1* = Johnson, 1972a; *2* = Johnson, 1972b; *3* = Lonsdale et al., 1972; *4* = Malfait, 1974; *5* = Lonsdale and Malfait, 1974; *6* = Lonsdale, 1975a; *7* = Lonsdale, 1974; *8* = Normark and Spiess, 1976; *9* = Hollister et al., 1974; *10* = Hollister et al., 1976; *11* = Johnson and Lonsdale, 1976; *12* = Lonsdale, 1975b; *13* = Hagen and Lonsdale, 1976. *14* = Shor et al., 1976; *15* = Hollister et al., 1976; *16* = Lonsdale and Hollister, 1976; *17* = Lonsdale and Hollister, in prep.

transverse to the flow, with downstream slip-faces, though isolated dunes are sometimes aligned in rows parallel to the flow. Difficult to recognize on bottom photos, and seldom visible on surface-ship echograms: best observed with side-looking sonars.

(3) *Mud Waves*: regular, large-scale ridges of cohesive muddy sediment, molded by current-controlled differential deposition. The smaller examples are well displayed on side-looking sonar records, while the largest abyssal mud waves (with wavelengths of several kilometers and amplitudes of tens of meters) can be studied on surface-ship echograms.

(4) *Furrows*: erosional troughs up to several kilometers in length; and regularly spaced at intervals of 10—100 m. Best studied with side-looking sonar and carefully positioned bottom photographs.

We adopt Allen's (1970) definition of bedforms as "spatially periodic mounds and hollows fashioned at the sediment—fluid interface by the action of tangential fluid forces", and concentrate our discussion on these regularly spaced phenomena. We have also observed a number of other more isolated pieces of evidence relating to existence of substantial currents at the sea floor. These include the upstream scour crescents and lee-side sediment shadows seen at several sites in the North Atlantic (Table I), deeply eroded large-scale moats in the Line Islands archipelagic apron (Normark and Spiess, 1976), and extreme examples of the stripping away of all sediment cover down to some particularly resistant horizon as at the Carnegie Ridge and Ecuador Trench sites. One further class of features, that is probably related to bottom currents, is exemplified by the lineated patterns of manganese nodules observed in the vicinity of the Samoan Passage (Lonsdale, 1974, fig.20) and, on a smaller scale, on the Carnegie Ridge (Lonsdale and Malfait, 1974, fig.7). These may be caused either by differential deposition of nodules and sorting during rolling, or by uneven deposition or erosion of a thin blanket of fine-grained sediment.

RIPPLES

There is a wonderful variety of ripple marks in the deep ocean. Early photographic work on the crests and flanks of seamounts (e.g., Heezen, 1959; Shipek, 1962) established the surprising abundance of symmetric sand ripples, apparently formed by oscillatory wave motion, at depths of 1—3 km. The commonest rippled sediment is winnowed foraminiferal sand, composed of hollow, low-density calcite spheres that can move as a saltating bed-load (and hence form ripples) at current speeds as low as 15—20 cm/sec. We have investigated such ripples on the flanks of Horizon Guyot (Lonsdale et al., 1972), where extensive trains show all the typical features of wave ripples; i.e., they are commonly symmetric, with a low-ripple index (length/height), and parallel straight crests with either tuning fork or open junctures. Extensive photography with the transponder-navigated deep tow established the patchiness and abrupt boundaries of rippled sand on Horizon Guyot, characteristics common to most seamount occurrences (e.g., Laughton et al., 1960).

We measured oscillatory currents of the internal tide which attained speeds of up to 17 cm/sec, 12 m above rippled sea floor (Lonsdale et al., 1972), and therefore ascribe the formation of wave ripples on seamounts to the action of internal waves impinging on slopes. This process may operate almost continuously, maintaining the observed sharpness of ripple crests and erasing the tracks and trails of benthic organisms. Some alternative explanations of their origin (e.g., by the passage of a tsunami) imply a much more episodic formation.

Laughton (1963) stated that no ripples had been found on the floor of the ocean basins, and Heezen and Hollister (1964) admitted that they were very rare there, except in such constrictions as Drake Passage. Subsequently, extensive photographic work by USNS "Eltanin" in the Southern Ocean discovered abundant, well-formed sand ripples beneath the circumpolar current at depths as great as 4500 m (e.g., Jacobs et al., 1970), and similar trains of ripples have been described from comparable depths on the New England continental rise that are swept by a western boundary current (Zimmerman, 1972). In all cases, these deep ripples on gentle slopes are highly asymmetric current ripples, evidently propelled by thermohaline currents of the general circulation. We have recently deep towed extensive patches of such ripples in the northeastern Atlantic, where they occur at depths of 2400—2900 m near both margins of Rockall Trough (Lonsdale and Hollister, in prep.), and at 3000 m on the lower flanks of Hatton Drift (Hollister et al., 1976). At the latter site, we measured fairly steady near-bottom currents, flowing northwest with an average speed of 14.9 cm/sec, and a maximum speed (during the one week of record) of 22 cm/sec. These speeds are close to the threshold for sand-ripple development, and the ripples photographed (Fig.2b) are the fairly straight, long-crested variety diagnostic of "low energy" environments (Harms, 1969). We have found high-energy linguoid ripples with discontinuous crests, indicative of faster bottom currents (Harms, 1969), in narrow passages where the abyssal circulation is accelerated, for example in the Ecuador Trench where average measured currents at 2900 m exceed 30 cm/sec (Hagen and Lonsdale, 1976), and in a channel across the Carnegie Ridge (Lonsdale and Malfait, 1974). At the latter site some of the high-energy ripples are superimposed on large-scale sand waves, which replace sand ripples at higher current velocities, if the supply of sand is adequate.

The dynamics of abyssal sand ripples are well understood, because of their similarity to shallow-water forms, and most of the uncertainty about them concerns the nature and origin of the forcing flow. Less is known about abyssal mud ripples, which occur where thermohaline currents flow over the fine-grained sediment more typical of depths below 4.5 km. Transverse mud ripples (Figs.2d and 6c) are usually small asymmetric current ripples (wavelength = 10—30 cm, amplitude 1—3 cm) with rounded crest profiles and no avalanching slip face. Another type of ripple formed in silty clays is sharp-crested and appears to be a longitudinal bedform. We have surveyed an extensive patch of these ripples on the Blake-Bahama Abyssal Plain

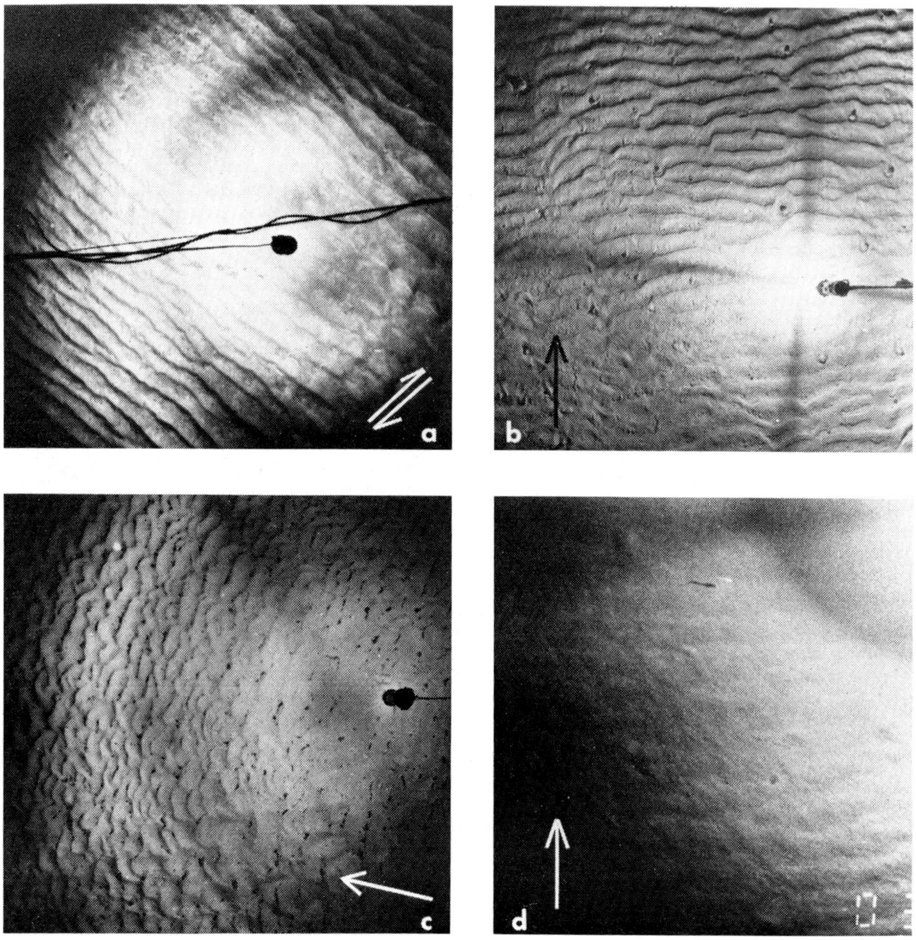

Fig.2. Deep-tow photographs of ripples. Arrows show current direction, and are 1 m long. a. Symmetric wave ripples. Horizon Guyot; water depth 1700 m; max. measured current 17 cm/sec. b. Long-crested low-energy current ripples. Hatton Drift; 3000 m; 22 cm/sec. c. High-energy linguoid ripples. North Carnegie Ridge; 2625 m; 19 cm/sec (probably formed by faster episodic flows). d. Mud ripples. Charlie-Gibbs Transform fault; 3650 m; 21 cm/sec.

(Fig.3b), where they have similar dimensions (wavelength of 1—2 m, amplitude of about 10 cm) to those described on the Mozambique Abyssal Plain by Heezen and Hollister (1964). These large ripples are just discernible on side-looking sonar records (Fig.3a). Longitudinal mud ripples have been described from shallow marine environments (Van Straaten, 1951), but their relationship to any particular flow regime is not understood. Their occurrence on parts of the deep-sea floor may indicate a small-scale secondary circulation which causes regular cross-stream variations within the benthic boundary layer.

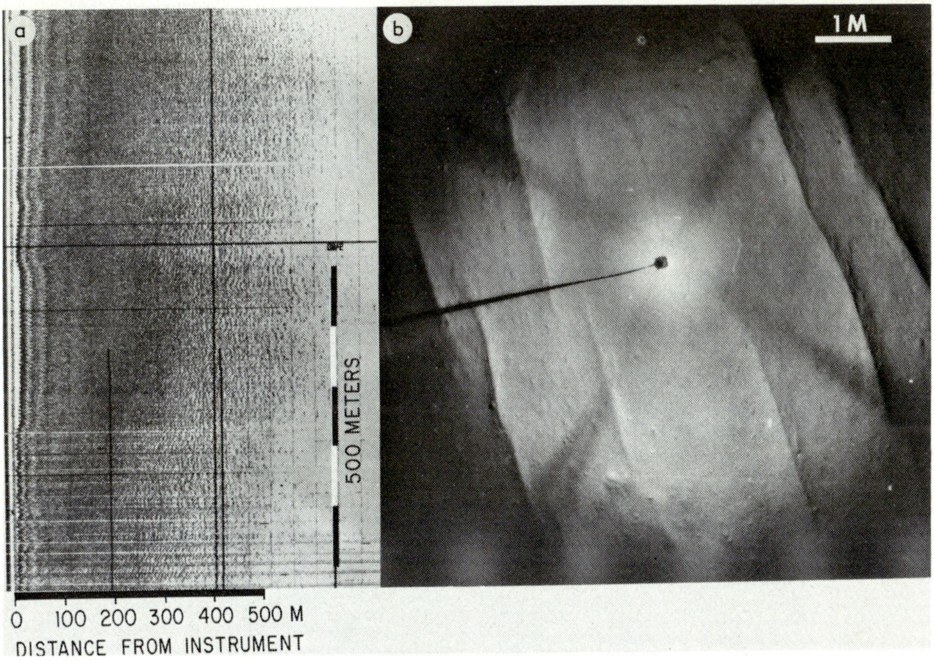

Fig.3. Large longitudinal mud ripples on the Blake-Bahama Abyssal Plain near 28° 34′N, 75° 25′W; water depth 4980 m. a. Side-looking sonar record, with ripples just visible as discontinuous wavy lines almost parallel to the track. b. Bottom photograph.

SAND WAVES

A field of large-scale waves of foraminiferal sand, discovered at a depth of 2650 m in a valley on the Carnegie Ridge (eastern equatorial Pacific) by side-looking sonars (Fig.4a) has been described in detail by Lonsdale and Malfait (1974). They also discuss the difficulty of recognizing these large features (with average wavelengths of 20 m) on conventional narrow-angle bottom photographs, and the impossibility of seeing them on surface-ship echograms because of their low amplitudes (<1 m). The slip faces of similar sand waves have been recognized on bottom photographs from 4500-m depths in Drake Passage, and we have subsequently found their characteristic patterns on side-looking sonar records from the northeast Atlantic (Table I), but their global distribution is poorly known.

During 1974 on Expedition "Cocotow" (Lonsdale, 1975b) we resurveyed the field of sand waves on the Carnegie Ridge that had been discovered in 1972. After an interval of 30 months, individual dunes could still be recognized in the same relative position, indicating little, if any, migration during that interval. Lenses of light detrital material ("organic fluff") have accumulated in the angle at the foot of the slip faces (Fig.4b), suggesting a lack of recent avalanching there. A current meter deployed on the dune field for

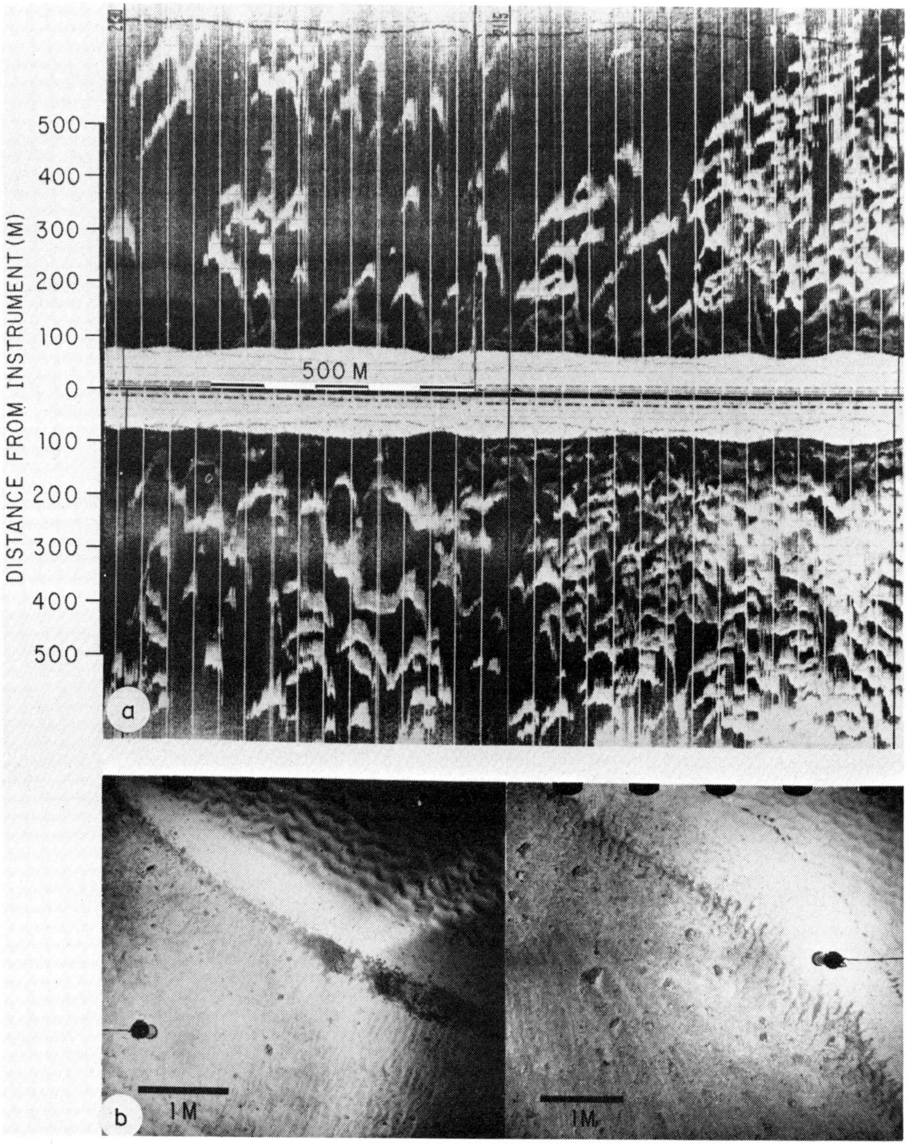

Fig.4. Sand waves on the Carnegie Ridge (water depth 2650 m). a. Pair of side-looking sonar records: sand waves are the light patches of lower backscattering, while manganese-encrusted substrate appears dark. b. Two photographs of slip faces. Note dark organic fluff collected in its lee, and small longitudinal ripples downstream.

16 days recorded a slow current that was flowing up-valley as often as down-valley (the direction of dune motion indicated by the orientation of their slip faces), with maximum speeds of less than 20 cm/sec, compared to the 30 cm/sec + required for dune migration. The geologic and hydrographic

data suggest that these sand waves migrate episodically, probably during spill-over of dense water across the Carnegie Ridge (Lonsdale and Malfait, 1974), which evidently occurs at intervals of several years or more.

The field of sand dunes on the Carnegie Ridge includes isolated "barchans", which are arranged in down-current rows just as subaerial desert barchans often are. Some of the abyssal barchan rows are fed by thin sand streamers, narrow longitudinal bands of sand which extend out from sediment reservoirs across a bare manganese pavement (Lonsdale and Malfait, 1974). These are the only longitudinal patterns of sand that we have so far encountered in the deep sea, though several types of large-scale longitudinal sand waves have been described from shallow waters of the continental shelf (e.g., Caston, 1972; Belderson and Stride, 1972).

MUD WAVES

Less is known about the muddy counterparts of sand waves. Johnson and Lonsdale (1976) describe regular elongate hummocks, with wavelengths of 10—20 m, amplitudes of 1—2 m and crests hundreds of meters long that are composed of fine-grained hemipelagic mud on the New England continental rise. Though well-displayed on side-looking sonar records (Fig.5), they are

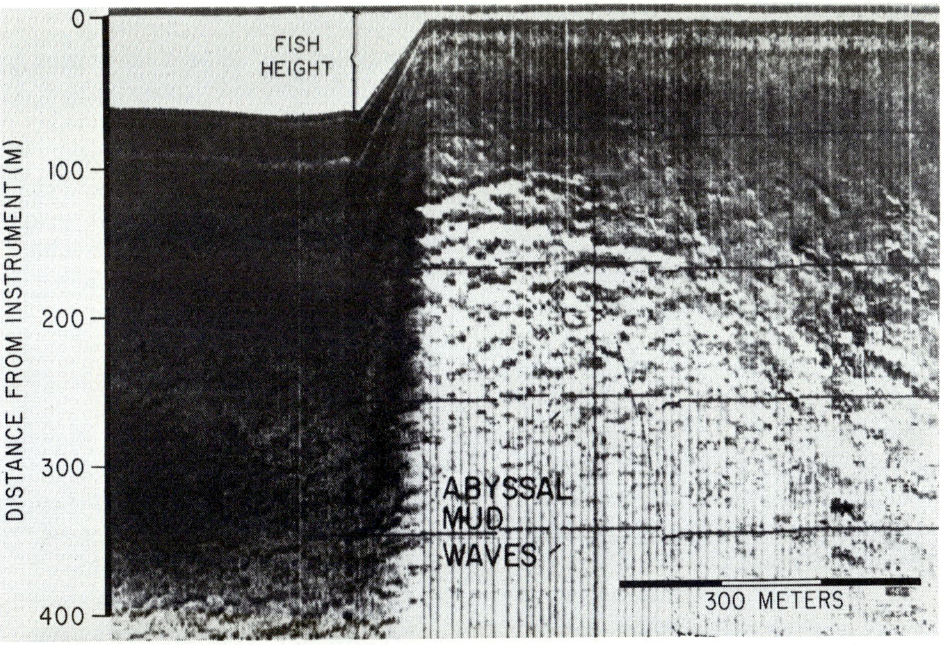

Fig.5. Side-looking sonar record of mud waves on the New England continental rise, in the moat adjacent to Mytilus Seamount (from Johnson and Lonsdale, 1976). Water depth 3700 m.

not resolved on surface-ship echograms, and our attempts to photograph them were thwarted by the very high turbidity of bottom water at this site. Johnson and Lonsdale (1976) deduce that they are depositional bedforms: they are aligned parallel to the contours, and probably parallel to the current that molded them (although an adjacent current meter recorded an alternation of along-slope and up-slope currents, in neither case exceeding 10 cm/sec).

Giant symmetrical mud waves, readily resolved by surface-ship profiles, are common on many continental rises. A deep-tow survey on the Blake-Bahama Outer Ridge (Hollister et al., 1974) straddled several of these features, which there have a wavelength of 2—3 km, and amplitudes of 40—50 m. They were oriented at about 35° to the slow (5—10 cm/sec) bottom currents measured over them. Records from the deep-tow 4-kHz seismic profiler across these giant mud waves (and similar features that we have examined on Feni Ridge in Rockall Trough) suggest that they have been formed by continuing differential deposition, with subbottom layers thickening over their crests and thinning in the troughs.

FURROWS

The most spectacular, and perhaps the most important, erosional bedform of the deep sea floor is the abyssal furrow. These are regularly spaced grooves, 1—20 m deep, that extend parallel to the flow direction for several kilometers. Although the larger and deeper examples do affect surface-ship echograms (causing "mushy" returns and a confusion of hyperbolic echoes), an adequate description of these phenomena requires deeply towed side-looking sonars and carefully positioned bottom photographs. Because the spacing of furrows is very much greater than their width, random photographs taken in conventional manner from a drifting ship failed to discover them even in regions such as the Bahama Outer Ridge that subsequent deep-tow surveys showed to be densely furrowed. Indeed, even the shallow-water analogues of these bedforms, occurring on estuarine mud flats, were first discovered with side-looking sonar techniques (Dyer, 1970).

The deep-tow system first encountered abyssal furrows at a depth of 5800 m near the exit of the Samoan Passage in the southwest Pacific, where they are incised into patches of reworked calcareous nannofossil ooze (Lonsdale et al., 1973; Lonsdale, 1974). These examples (Fig.6a) are about a meter deep, spaced an average of 30 m apart, and are slightly curved to conform with the local contours. Individual furrows join with characteristic tuning-fork junctions. Measured bottom currents adjacent to one patch of furrows averaged only 2.5 cm/sec, but less than 30 km upstream (within the Samoan Passage) average velocities exceed 17 cm/sec, and there is some geologic evidence that the furrows are relics from a period when the region of fast flow was more extensive (Lonsdale, 1974)

During 1973 we explored similar erosional furrows over an extensive area on the Bahama Outer Ridge (Hollister et al., 1974a). Near the crest of this

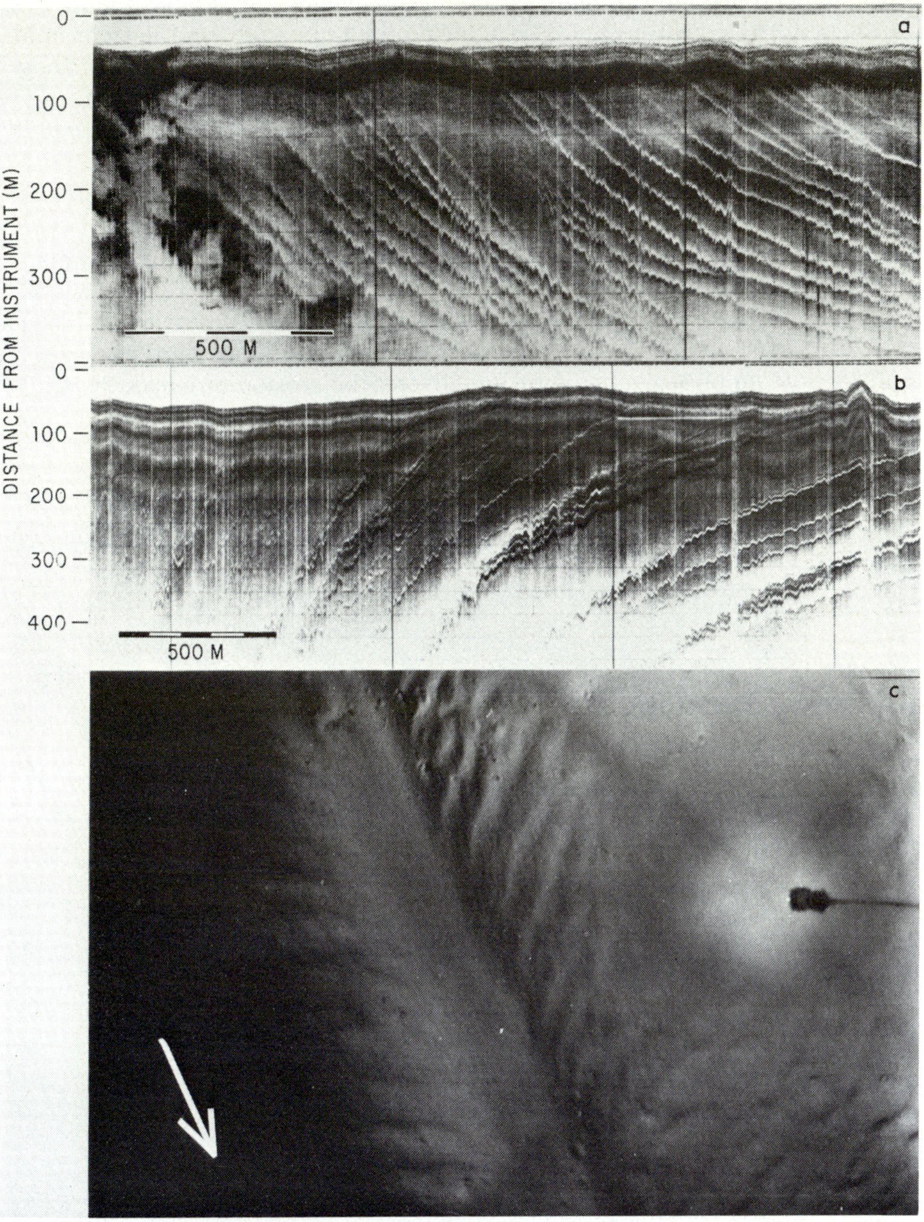

Fig.6. Abyssal furrows in the Pacific and Atlantic. a. Side-looking sonar record of a patch of furrowed calcareous ooze near the exit of the Samoan Passage at 7° 15′S, 168° 28′W. Water depth 5760 m. Note tuning-fork branches at upper right. Dark areas of high backscatter at left are local patches of manganese nodules. b. Side-looking sonar record of furrows in hemipelagic silty clay on the Blake-Bahama Outer Ridge at 28° 17′N, 74° 24′W. Water depth 4760 m. c. Photograph of a Blake-Bahama Outer Ridge furrow. Arrow shows current direction measured nearby, and is 1 m long. Note that the apparent curvature of furrows on the side-looking sonar records is caused by the curved path of the towed instrument package; the small-scale zigzag nature of the (white) furrow shadows is caused by minor pitching and yawing of the instrument.

huge depositional landform are erosional furrows of comparable size to those in the southwest Pacific (Fig.6b), superimposed obliquely on the giant mud waves described above. Photographs of the grooves (Fig.6c) show that they have convex, ripple-covered sides, and flat floors that may reflect a lithologic control on their morphology. The sediment into which they are cut is hemipelagic silty mud.

Another deep-tow survey, at the western foot of the Bahama Outer Ridge near its contact with the Bahama Abyssal Plain, revealed that the cause of hyperbolic echoes on surface-ship echograms was a train of much deeper (10—20 m), broader (50—150 m) furrows, spaced 50—200 m apart (Fig.7a). Their erosional nature is demonstrated by the truncation of internal reflectors (mapped by the deep-tow 4-kHz profiler system) on their walls, and by bottom photographs showing differential erosion of ledges there (Fig.7b). However, the maximum near-bottom current speed recorded over

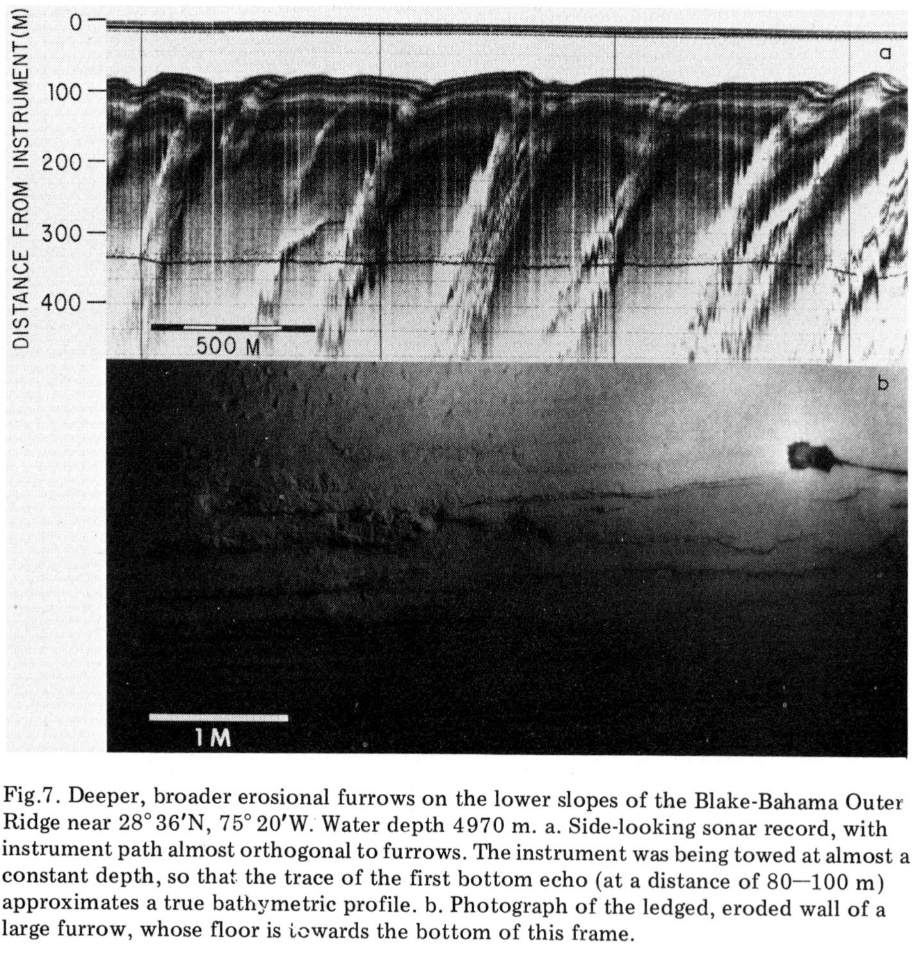

Fig.7. Deeper, broader erosional furrows on the lower slopes of the Blake-Bahama Outer Ridge near 28° 36′N, 75° 20′W. Water depth 4970 m. a. Side-looking sonar record, with instrument path almost orthogonal to furrows. The instrument was being towed at almost a constant depth, so that the trace of the first bottom echo (at a distance of 80—100 m) approximates a true bathymetric profile. b. Photograph of the ledged, eroded wall of a large furrow, whose floor is towards the bottom of this frame.

the field (though not within a furrow) was only 8 cm/sec (Hollister et al., 1974).

The most notable features of abyssal furrows, particularly the smaller variety (Fig.6), are their great length, narrow width, and regular spacing. In each of the three furrowed sites the measured currents have been parallel to the bedforms (and directed into the acute angle of tuning-fork junctions), and have been much steadier in direction than most relatively slow abyssal currents, which are generally more perturbed by tidal oscillations. We infer from the furrow morphology that there are narrow threads of high-velocity current moving along the furrows (rippling and eroding their walls, see Fig.6c, and bowling along large masses of organic debris, see Hollister et al., 1974), separated by broad bands of much slower velocity (e.g., the 8 cm/sec currents we measured on the Blake-Bahama Outer Ridge) which have done little more than slightly smooth the surface of inter-furrow areas. It is possible that the origin of this cross-stream variation in the speed of the current impinging on the bottom is a helical secondary circulation in the low stability benthic boundary layer, with the spacing of furrows perhaps controlled by the spacing of helical vortices. This type of circulation in the atmospheric boundary layer is held responsible for a variety of linear natural phenomena; e.g., the longitudinal sand dunes of trade-wind deserts (Hanna, 1969). However, there are difficulties with such an interpretation of the abyssal furrows, especially in explaining why the strips affected by high-speed currents are so narrow. In any event, the geologic evidence indicates a surprising and unsuspected hydrodynamic regime in the boundary layers of some major thermohaline currents, deserving further study with closely and precisely spaced hydrographic observations, and proper hydrodynamic analyses. If the cross-current variations in near-bottom speeds are as extreme as our indirect observations suggest, then it is clear that the standard widely-spaced measurements of near-bottom currents may yield an inaccurate, or at least very incomplete, estimate of the average speed, transport, and geologic potential of major thermohaline flows.

DISTRIBUTION OF ABYSSAL BEDFORMS

Necessary conditions for the formation of abyssal bedforms are the availability of sufficient sediment and the presence of a bottom current competent to move it and persistent enough to mold it. The type of bedform produced is determined by the properties of the sediment, especially its grain size and cohesion, and of the bottom current, especially its speed and steadiness.

Most of the deep-sea floor is reputed to be swept by very slow (<2 cm/sec) thermohaline currents, which transport cold dense deep and bottom water from its polar sources to all other parts of the ocean, where it moves slowly upwards. Our surveys (Table I) are concentrated on sites of usually fast flow, and these are of two general types: near sills of narrow passages where constricted thermohaline flows are forced to accelerate (e.g., the Samoan Passage) and at the ocean basin boundaries, both western (e.g., Blake-Bahama Outer

Ridge) and eastern (e.g., Hatton Drift), where there is an intensification of the general circulation, caused by the Earth's rotation. The density-driven flows seek the deepest routes throughout the oceans, and the thermohaline circulation at mid-depths (1—3 km) and in enclosed deeps is usually slower. Important exceptions to this generalization occur near the sources of dense polar water (e.g., in the Southern Ocean, where the Circumpolar Current extends from the surface to the ocean basin floor), and near sills into or out of large basins whose water has a short residence time (e.g., Panama Basin, Norwegian Sea). Tidal currents are ubiquitous in the deep ocean, but their speeds and geological effectiveness are poorly understood: as far as is known they are seldom fast enough to entrain sediment, except on sloping terrain at mid-depths.

The deeper parts of the ocean basins are covered with fine-grained sediments with such a small coarse fraction that massive winnowing is required to form a layer of cohesionless sand. There are photographs and cross-bedding evidence of deep-sea ripples composed of volcanic ash and mineral grains, but winnowed foraminiferal ooze is by far the commonest substrate of sand ripples and dunes, and these bedforms are effectively restricted to depths shoaler than the calcium carbonate compensation depth (on average 4.5 km in the Pacific and 5 km in the Atlantic, see Berger and Winterer, 1974). Many of the faster thermohaline currents, especially of Antarctic Bottom Water in the Pacific Ocean, are concentrated below these depths, and therefore lack the bed of sandy current ripples and sand waves found beneath the Circumpolar Current and shallow parts of the North Atlantic Deep Water currents (e.g., Fig.2b). Current ripples and waves of foraminiferal sand are found at mid-depth basin inflows and outflows, though in several examples we have examined (e.g., the Ecuador Trench) the currents are so fast that they have scoured the passages almost free of unconsolidated sediments, and the bedforms are starved (forming isolated piles of sand migrating over a bare rock pavement). Passages with fast, episodic inflow or outflow may be favored sites for large-scale sand waves, which require fast currents to form, and yet may be swept too far from their limited source of sand if the current is persistent: the best examples we know of (on the Carnegie Ridge and in Drake Passage) are also beneath highly productive surface waters, and receive a faster-than-average supply of foraminiferal tests.

Wave-formed sand ripples are widespread on ooze-covered slopes of seamounts and ridges at mid-depths (Heezen and Hollister, 1964, report that over three-quarters of the photographs of these environments show "dramatic current evidence"). An explanation of their local distribution awaits better understanding of the internal waves we believe to be responsible.

The distribution of abyssal furrows is poorly known, because of the small proportion of the sea floor that has been surveyed with deeply towed side-looking sonars. However, wide areas on the North American continental rise and elsewhere in the Atlantic return a confusion of hyperbolic echoes similar to the surface manifestations of the larger furrows on the Blake-Bahama Outer Ridge. The morphology of abyssal furrows indicates that very steady bottom currents are required for their formation: if the direction of erosive flow deviated by more than a few degrees they could not attain such a long, linear shape. Tidal oscillations are an important cause of unsteadiness in most deep-sea currents, but in this part of the northwest Atlantic the predicted semi-diurnal tidal currents are very low (Pekeris and Accad, 1969, fig.10), and they were barely resolved by our near-bottom current meters. At Feni Ridge, in the northeastern Atlantic, our current meters revealed bottom currents with a large tidal oscillation (Lonsdale and Hollister, in prep.), and though there is evidence of significant erosion and surface-ship echograms show hyperbolic echoes we did not discover any comparable regular furrows.

The existence of erosional bedforms in the deep sea demonstrates not only the power of ocean-bottom currents, but also their changeability. For a current to actively erode a bed that it had previously deposited (as on the Blake-Bahama Outer Ridge, where furrows incise Pleistocene contourites) there must have been a change in speed or turbidity. In some instances other interpretations of the geological observations are possible, as at the Ecuador Trench, where a high-speed axial current erodes pelagic carbonates deposited in tranquil water, and delivered to the trench by plate motion (Lonsdale, in prep.). Major thermohaline currents that have unremittingly flowed over the same bed for millions of years (e.g., the western boundary current of Antarctic Bottom Water in the southwest Pacific) are generally floored with a pavement of manganese nodules, which are indicative of current-impeded deposition, and themselves impede further erosion.

ACKNOWLEDGEMENTS

Several geologists have helped collect and interpret deep-tow data on bedforms, especially R. Flood, C. Hollister, D. Johnson, B. Malfait, and W. Normark. We also acknowledge the enthusiastic support of the deep-tow engineers, diverse scientific parties, and the crews, officers, and captains of Scripps Institution of Oceanography and Woods Hole Oceanographic Institution vessels.

Our work has been supported by grants from NSF (principally NSF GA 31377X, GA 41358, DES 74—03690 and DES 74—20396) and various ONR contracts.

REFERENCES

Allen, J.R.L., 1968. On the character and classification of bedforms. Geol. Mijnbouw, 47: 173—185.

Allen, J.R.L., 1970. Physical Processes of Sedimentation. Allen and Unwin, London, 248 pp.

Belderson, R.H., Kenyon, N.H., Stride, A.H. and Stubbs, A.R., 1972. Sonographs of the Sea Floor. A Picture Atlas. Elsevier, Amsterdam, 185 pp.

Berger, W.H. and Winterer, E.L., 1974. Plate stratigraphy and the fluctuating carbonate line. Spec. Publ. Int. Assoc. Sedimentol., 1: 11—18.

Boegeman, D.E., Miller, G.J. and Normark, W.R., 1972. Precise positioning for near-bottom equipment using a relay transponder. Mar. Geophys. Res., 1: 381—396.

Caston, V.N.D., 1972. Linear sand banks in the southern North Sea. Sedimentology, 18: 63—78.
Dyer, K.R., 1970. Linear erosional furrows in Southampton water. Nature, 225: 56—58.
Egloff, J., 1972. Morphology of the ocean basin seaward of northwest Africa: Canary Islands to Monrovia, Liberia. Am. Assoc. Pet. Geol., 56: 694—706.
Ewing, M., Eittreim, S.L., Ewing, J.I. and Le Pichon, X., 1971. Sediment transport and distribution in the Argentine Basin. 3. Nepheloid layer and processes of sedimentation. Phys. Chem. Earth, 8: 51—77.
Fox, P.J., Heezen, B.C. and Harrian, A.M., 1968. Abyssal anti-dunes. Nature, 220: 470—472.
Goodell, H.G., Meylan, M.A. and Grant, B., 1971. Ferromanganese deposits of the South Pacific Ocean, Drake Passage, and Scotia Sea. In: J.L. Reid (Editor), Antarctic Oceanology. Antarct. Res. Ser., 15: 27—92.
Hagen, R. and Lonsdale, P.F., 1976. Inflow of bottom water to the Panama Basin. EOS, Trans. Am. Geophys. Union, 57: 260 (abstract).
Hanna, S.R., 1969. The formation of longitudinal sand dunes by large helical eddies in the atmosphere. J. Appl. Meteorol., 8: 874—883.
Harms, J.C., 1969. Hydraulic significance of some sand ripples. Geol. Soc. Am. Bull., 80: 363—396.
Heezen, B.C., 1959. Dynamic processes of abyssal sedimentation: Erosion, transportation and redeposition on the deep sea floor. J. Geophys. R. Astron. Soc., 2: 142—163.
Heezen, B.C. and Hollister, C.D., 1964. Deep-sea current evidence from abyssal sediments. Mar. Geol., 1: 141—174.
Heezen, B.C. and Hollister, C.D., 1971. The face of the Deep. Oxford University Press, New York, N.Y., 659 pp.
Hollister, C.D. and Elder, R.B., 1969. Contour currents in the Weddell Sea. Deep-Sea Res., 16: 99—101.
Hollister, C.D. and Heezen, B.C., 1971. Geological effects of ocean bottom currents: western North Atlantic. In: A.L. Gordon (Editor), Studies in Physical Oceanography, 2. Gordon and Breach, New York, N.Y., pp. 37—66.
Hollister, C.D., Flood, R.D., Johnson, D.A., Lonsdale, P.F. and Southard, J.B., 1974. Abyssal furrows and hyperbolic echo traces on the Bahama Outer Ridge. Geology, 2: 395—400.
Hollister, C.D., Gardner, W.D., Lonsdale, P.F. and Spencer, D.W., 1976. New evidence for northward-flowing bottom water along the Hatton sediment drift, eastern North Atlantic. EOS, Trans. Am. Geophys. Union, 57: 261 (abstract).
Jacobs, S.S., Bruchhausen, P.M. and Bauer, E.B., 1970. Eltanin Reports, Cruises 32—36, 1968. Lamont-Doherty Geological Observatory, Palisades, N.Y. 460 pp.
Johnson, D.A., 1972a. Ocean floor erosion in the equatorial Pacific. Geol. Soc. Am. Bull., 83: 3121—3144.
Johnson, D.A., 1972b. Eastward-flowing bottom currents along the Clipperton Fracture Zone. Deep-Sea Res., 19: 253—257.
Johnson, D.A. and Lonsdale, P.F., 1976. Erosion and sedimentation around Mytilus Seamount, New England Continental Rise. Deep-Sea Res., in press.
Laughton, A.S., 1963. Microtopography. In: M.N. Hill (Editor), The Sea, 3. Wiley, New York, N.Y., pp. 437—472.
Laughton, A.S., Hill, M.N. and Allen, T.D., 1960. Geophysical investigation of a seamount 130 miles north of Madeira. Deep-Sea Res., 7: 117—141.
Lonsdale, P.F., 1974. Abyssal Geomorphology of a Depositional Environment at the Exit of the Samoan Passage. Thesis, Univ. of Calif., San Diego, Calif., 106 pp. (unpublished).
Lonsdale, P.F., 1975a. Sediment and tectonic modification of the Samoan archipelagic apron. Am. Assoc. Pet. Geol. Bull., 59: 780—798.
Lonsdale, P.F., 1975b. Detailed abyssal sedimentation studies in the Panama Basin. Cruise Report, "Cocotow" Legs 2b and 3. SIO Ref. 75—4.

Lonsdale, P.F. and Hollister, C.D., 1976. Cut-off of an abyssal meander on the Icelandic insular rise. EOS, Trans. Am. Geophys. Union, 57: 269 (abstract).

Lonsdale, P.F. and Hollister, C.D., in prep. A near-bottom traverse of Rockall Trough: hydrographic and geologic inferences.

Lonsdale, P.F. and Malfait, B.T., 1974. Abyssal dunes of forminiferal sand on the Carnegie Ridge. Geol. Soc. Am. Bull., 85: 1697—1712.

Lonsdale, P.F., Normark, W.R. and Newman, W.A., 1972. Erosion and sedimentation on Horizon Guyot. Geol. Soc. Am. Bull., 83: 289—316.

Lonsdale, P.F., Spiess, F.N. and Mudie, J.D., 1973. Erosional furrows across the abyssal Pacific floor. EOS Trans. Am. Geophys. Union 54: 1110 (abstract).

Malfait, B.T., 1974. The Carnegie Ridge at 86° W: Structure, Sedimentation, and Near-Bottom observations. Thesis, Oregon State Univ., Corvallis, Ore. (unpublished).

Menard, H.W., 1952. Deep ripple marks in the sea. J. Sediment. Petrol., 22: 3—9.

Normark, W.R. and Spiess, F.N., 1976. Erosion on the Line Islands archipelagic apron: Effect of small-scale topographic relief. Geol. Soc. Am. Bull., 87: 286—296.

Pekeris, C.L. and Accad, Y., 1969. Solution of Laplace's equation for the M_2 tide in the world oceans. Philos. Trans. R. Soc. London, Ser. A, 265: 413—436.

Potter, P.E. and Pettijohn, F.J., 1963. Palaeocurrents and Basin Analysis. Springer, Berlin, 296 pp.

Reineck, H.E. and Singh, I.B., 1973. Depositional Sedimentary Environments. Springer. Berlin, 439 pp.

Rona, P.A., 1969. Linear "lower continental rise hills" off Cape Hatteras. J. Sediment. Petrol., 39: 1132—1141.

Schick, G.B.E., Isaacs, J.D. and Sessions, M.H., 1968. Autonomous instruments in oceanographic research. In: Marine Sciences Instrumentation. Plenum, New York, N.Y., 4: 203—230.

Shipek, C., 1962. Photographic survey of sea floor on southwest slope of Eniwetok Atoll. Geol. Soc. Am. Bull., 73: 805—812.

Shor, A.N., Lonsdale, P.F., Hollister, C.D. and Spencer, D.W., 1976. Structure and sediment drift morphology in the Charlie-GibbsTransform Fault. EOS, Trans. Am. Geophys. Union, 57: 269 (abstract).

Spiess, F.N. and Mudie, J.D., 1970. Small-scale topographic and magnetic features. In: A.E. Maxwell (Editor), The Sea, 4. Wiley, New York, N.Y., pp. 205—250.

Spiess, F.N. and Tyce, R.C., 1973. Marine Physical Laboratory deep tow instrumentation system. SIO Ref. 73—4: 37 pp.

Van Straaten, L.M.J.V., 1951. Longitudinal ripple marks in mud and sand. J. Sediment. Petrol., 21: 47—54.

Zimmerman, H.B., 1972. Sediments of the New England Continental Rise. Geol. Soc. Am. Bull., 83: 3709—3272.

Marine Geology 23 (1977) 77—88

SEDIMENT TRANSPORT DOWN A SEAMOUNT FLANK BY A COMBINED CURRENT AND GRAVITY PROCESS

DANIEL JEAN STANLEY and PATRICK T. TAYLOR

Division of Sedimentology, Smithsonian Institution, Washington, D.C. (U.S.A.)
U.S. Naval Oceanographic Office, Washington, D.C. (U.S.A.)

(Received April 28, 1976)

ABSTRACT

Stanley, D.J. and Taylor, P.T., 1977. Sediment transport down a seamount flank by a combined current and gravity process. Mar. Geol., 23: 77—88.

A photographic study of the Gilliss Seamount reveals that gravity-induced processes, suggested by downslope-trending anastomosing tongues of varying-size material (volcanic rubble, organic debris, pelagic material), as well as currents, indicated by ripple marks and other features, prevail on its upper- to mid-flank sectors. Long-term current activity has dominated sediment dispersal on the lower slope to the base of this mount. A sediment transport model is proposed in which detritus is moved progressively downslope over a complex terrace-like topography by a sequence dominated alternately by current traction and gravity-influenced mechanisms. This mode of transport may be applicable to other submarine topographic highs.

INTRODUCTION

Seismic and sedimentological investigations in the different world oceans show that seamounts, guyots, knolls and ridges, most of which are topographic features of volcanic origin, are partially mantled by a sediment cover consisting in large part of pelagic detritus. It appears that ocean currents are most influential in the depositional pattern of suspensates which form acoustically transparent layers. It has been demonstrated that positive topographic features often act as obstacles to the circulating suspensate-bearing water masses and they localize depositional patterns in the abyssal areas by serving as collection foci for pelagic sediments.

The role of gravity transport processes, which may be locally important, has not been assessed completely. The purpose of the present study is to focus on the interaction between circulation and gravitational processes which are likely to modify sediment distribution on positive submarine topographic features.

The example used to illustrate this idea is the Gilliss Seamount (one of the Kelvin Seamount Chain) 666 km northeast of Bermuda (Fig.1). This twin-peaked mount is elongated in a NNE—SSW direction, and has a relief of

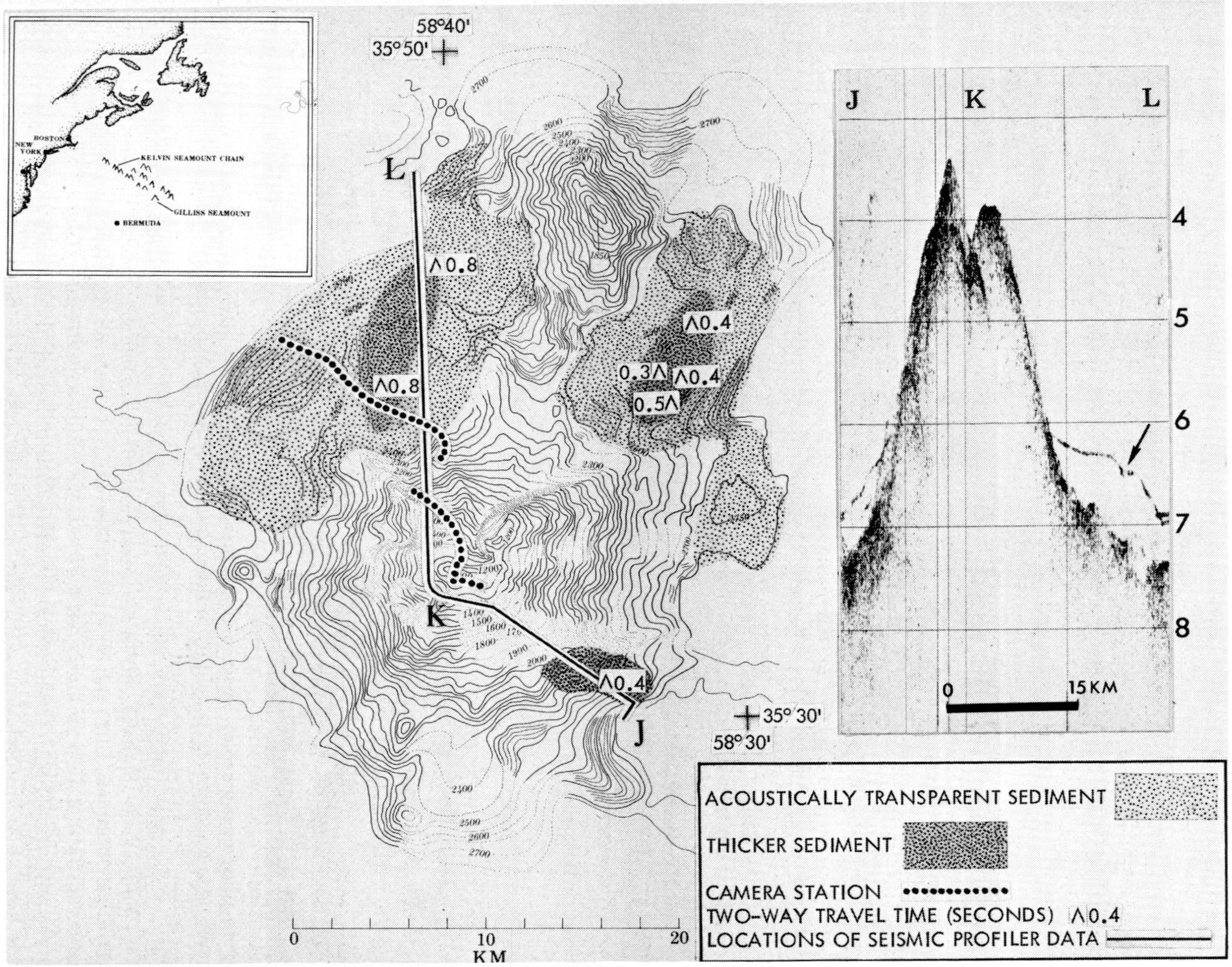

Fig.1. Photographic track lines on chart of the Gilliss Seamount. Bathymetry based on sonar array system data (modified from Taylor et al., 1975). Contour values are given in corrected fathoms, and contour interval is 50 fm above 2500 fm and 10 fm below this depth. Dashed lines indicate extrapolated contours between data tracks. Seismic profiler data (*J—K—L*) is presented on the right; vertical scale, two-way travel time in seconds.

about 3000 m between the highest peak and the surrounding Sohm Abyssal Plain. Detailed geological (bathymetry, sea-floor photography, cores, dredges) and geophysical descriptions of this seamount have been published (Cordell and Taylor, 1971; Taylor and Hekinian, 1971; Taylor et al., 1975). In the present study we use bottom photographs to ascertain the dominant transport mechanisms along one flank of this feature. Our results help us determine: (1) details of downslope sediment dispersal; (2) the importance of gravity in sediment transport; and (3) the influence of microtopography on the interaction between bottom currents and sediment gravity flow mechanisms. A gravity-current sediment transport model is proposed and considered in terms of analogous submarine features studied elsewhere.

OBSERVATIONS

Upper flank of Gilliss Seamount

The detailed bathymetric and seismic (sparker, 18 KJ) surveys of the Gilliss Seamount previously cited showed the extremely complex morphology of this feature and the variable distribution of the sediment cover mantling the volcanic core of the seamount (Fig.1). These data reveal that the thickest sections of acoustically transparent sediment, of probable Late Cretaceous to Quaternary age, are localized below the 4000-m (2200 fm) isobath in sectors NW and SE of the elongated axis of the twin peaks (thickness patterns in Fig.1). The surficial sediment is a tan-colored calcareous-sandy mud consisting of approximately equal amounts of silt plus clay fraction and planktonic foraminifera (Taylor et al., 1975). Volcanic debris is locally an important constituent. At shallower depths the mount appears free of sediment (see line *J—K—L* in Fig.1). However, photographic stations (dotted lines, Fig.1), intersecting the seismic line, reveal an apparently thin, irregularly distributed veneer of sediment above the 4000-m isobath.

The photographic track extending across the southern peak and toward the north and northwest down a steep flank shows the seamount crest locally mantled with apparently thin sediment as well as sediment-free igneous outcrops. The sequence of photographs indicates regions with poor depth of field (crestal zone, Fig.2B), due to particulate matter in suspension, alternating with areas of clear water (upper flank, Fig.2A). The sea-floor facies in this upper sector, where slopes vary from about 13° to 59°, are highly variable: (1) sediment-free volcanic exposures including massive lava flows and pillows (Fig.2A; see also Taylor et al., 1975, their fig.7); (2) very thinly sediment-mantled volcanic exposures (Fig.3); (3) and almost completely sediment-covered zones (Fig.2C, D). Current intensities are inferred by the following criteria, which are listed in order of relatively decreasing current activity: (1) sediment-free zones including oriented sessile-benthic organisms such as sea fans (Fig.2A); (2) concentrations of coarse granules and sand, sometimes molded into larger current-modified tongues (Taylor et al., 1975, their fig.5F); (3) coarse sand and granules molded into sharp

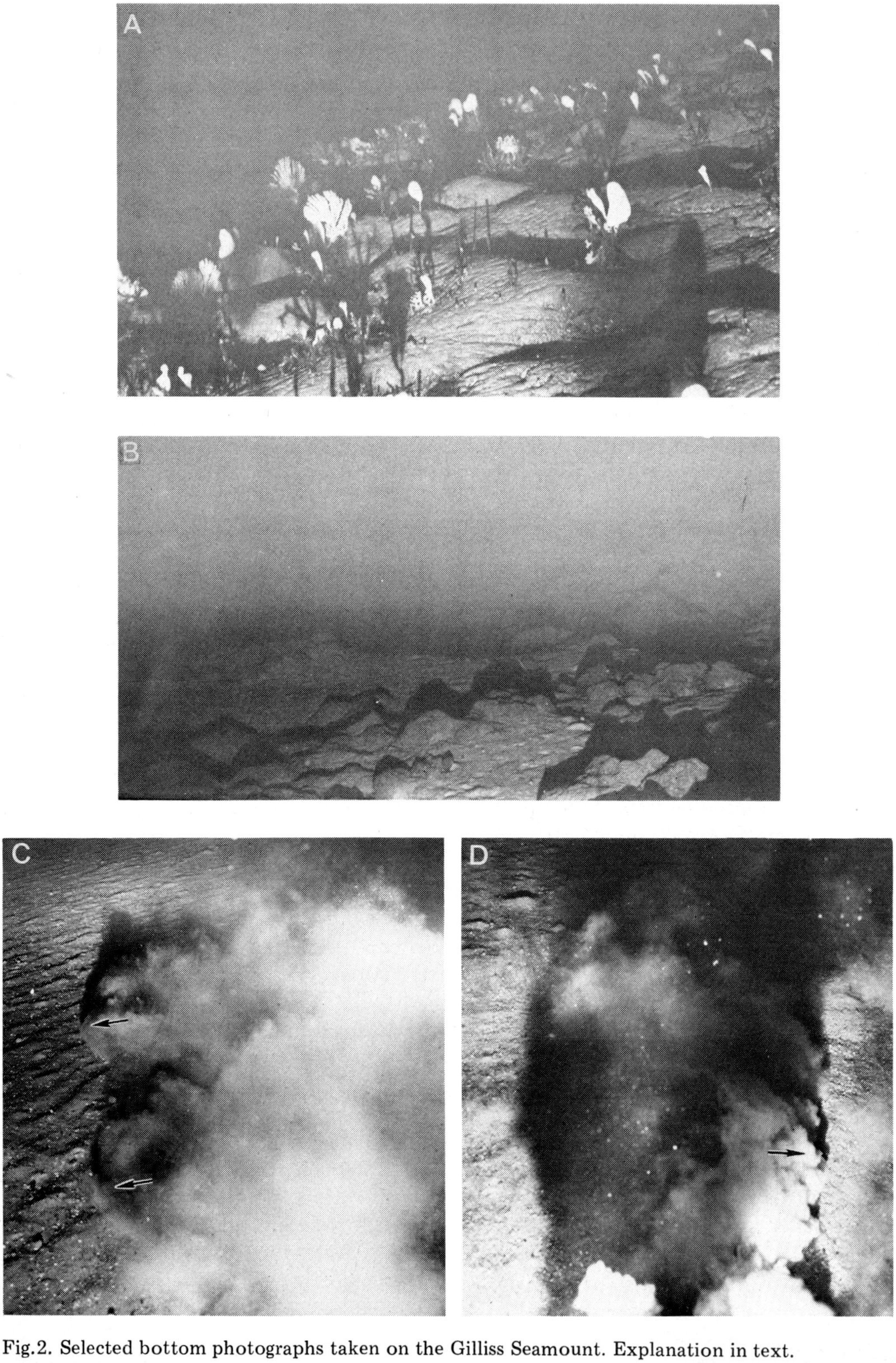

Fig.2. Selected bottom photographs taken on the Gilliss Seamount. Explanation in text.

Fig.3. Selected bottom photographs taken on the Gilliss Seamount. Explanation in text.

asymmetric ripples (Fig.3D, arrow); (4) poorly defined rippled sand and sandy silt (Fig.3C); (5) sediment scour around blocks and current tails (Taylor et al., 1975, their fig.5B); (6) sediment without clear evidence of current modification, frequently bioturbated (Fig.2B).

That currents in this sector can be intense is demonstrated by the presence of cleanly swept volcanic outcrops, coarse sand-granule lags, and sharp asymmetric ripple marks. Gorgonian sea fans and sea pens (Fig.2A) show an orientation normal to predominant current flow; it has been shown that this orientation is the result of their feeding mechanism whereby they filter nutrients from passing currents (Heezen and Hollister, 1971, pp.38–51; Grigg, 1972). It is clear from this visual data that erosion as well as transport by bottom currents is significant on the upper Gilliss flanks. Considering the above features in terms of the criteria presented by Heezen and Hollister

(1971, their figs.9.21, 9.24) we estimate the current velocities are locally in excess of 30 cm/sec. However, no information is known on the constancy of current intensity or direction.

In addition to the evidence for sediment movement by bottom currents, our interpretation of these photographs emphasize the importance of downslope movement due to gravity. The camera track reveals downslope-trending anastomosing streams of sand and gravel, varying from a few centimeters (Fig.3A) to a meter (Fig.3C) or more (Fig.3D) in width. These tongues include medium- to coarse-grained sand, gravel and not infrequently large volcanic debris including pillow fragments (Fig.3B). These tongues are interpreted as gravity-entrained "rivers of sand" apparently in a state of movement, probably by creep. One example (Fig.3D) reveals an absence of current markings in a channel axis compared to rippled sands on the margin; we interpret this to indicate relatively greater motion of sediment in the channel axis in contrast to the rippled channel margin. It is impossible, naturally, to illustrate sediment motion from still photographs, and our analyses are not, therefore, unique. The mechanisms involved for the downslope displacement also include rockfall (Dott, 1963) as well as debris flow and possibly sand flow processes (Middleton and Hampton, 1973). The frequent occurrence of criteria indicative of downslope processes, as evidenced by bottom photographs, illustrates the importance of sediment gravity flow in downslope movement of the sediment cover on the upper flank.

Lower flank of the Gilliss Seamount

A second photographic track (upper on Fig.1) traverses the lower sector of the seamount (about 4000—5000 m, or 2200—2700 fm), which has slopes of about 3° or less. This profile crosses the greatest thickness (about 650—700 m) of acoustically transparent sediment (Fig.1). As in the photographs of the upper flank, this sector shows surficial sediment textures ranging from mud (Fig.4A) to granule size (Fig.4D), some patchy volcanic exposures (Fig.4C) and evidence of current activity. Current tails (Fig.4B, arrows) and scour around igneous blocks and exposures (Fig.4C, arrow) are observed. As in the case of the upper slope, no indication of the duration or the intensity of current activity on the lower flank is ascertained.

In contrast to the upper flank, this portion of the seamount shows: (1) a more complete covering of the igneous exposures by sediment; (2) an apparently less intense current activity; and (3) a marked increase in bioturbation. The most intense and prevalent current activity inferred from an analysis of all the bottom photographs occurs in the vicinity of the step-like terrace notched in the transparent layer (Fig.1, seismic insert, arrow). This notch is probably the result of beveling by bottom-flowing currents, although slumping cannot be excluded. In their earlier study of the Gilliss Seamount, Taylor and Hekinian (1971, their figs.4 and 5) suggested that slumping might have produced such thickened sediment accumulations. McGregor et al. (1973), working in the Corner seamounts, also suggest that

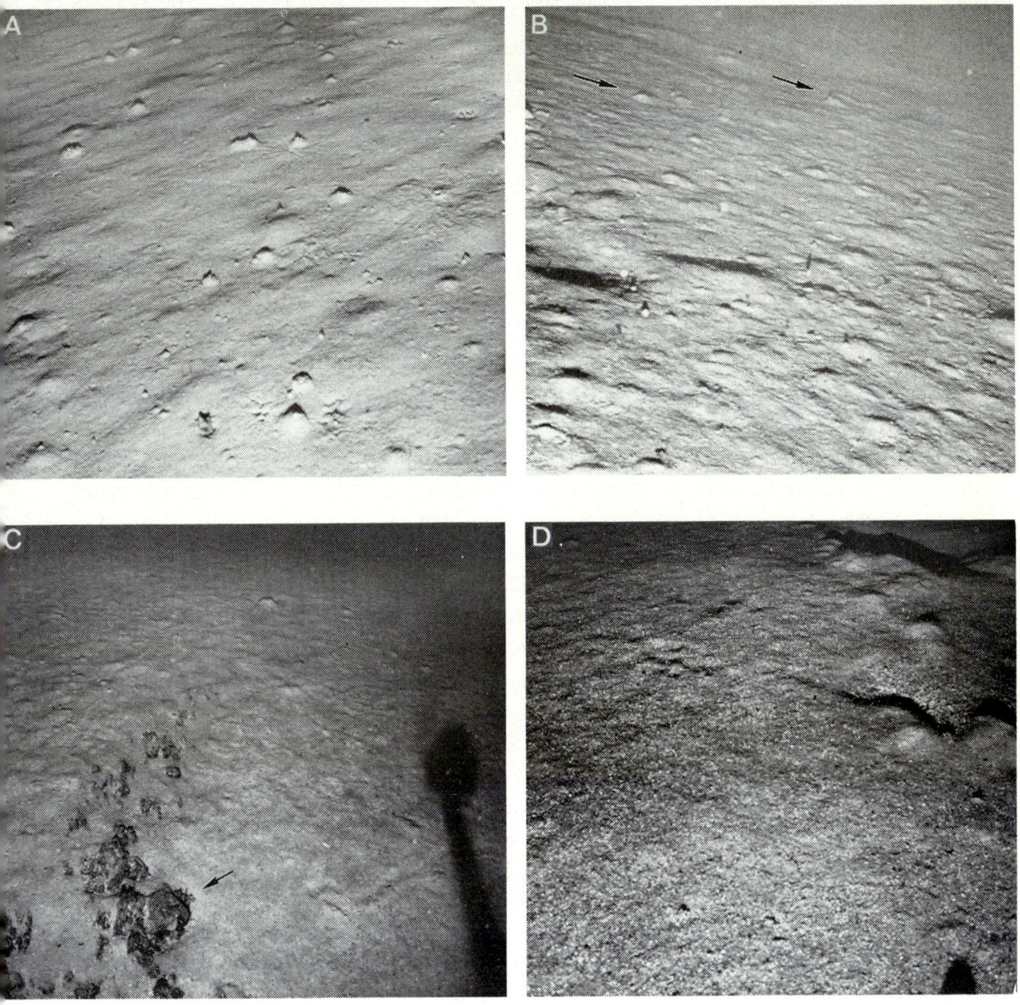

Fig.4. Selected bottom photographs taken on the lower flank of Gilliss Seamount. Explanation in text.

slump-like features occur where there are thicker sediment accumulations. In the case of the Gilliss, however, it should be emphasized that the occurrence of most intense current activity, as revealed from bottom photographs, is localized in the vicinity of this step-like feature.

In addition to all of the above-cited sediment gravity flow mechanisms, photographs also indicate the possible role of turbidity currents. Small turbidity currents were generated by the bottom contact weight used to activate the camera shutter, assuming that each photograph was taken during initial impact. Figs.2C and D illustrate two small artificially triggered flows, separated by at least 3 km on the upper part of the lower slope (greater than 3°); a similar process of turbidite generation has been observed and recorded on videotape by Stanley et al. (1972) in the Wilmington Canyon. This initiation of small sediment flows reflects the metastable condition of material on these slopes and the facility with which gravity can influence sediment transport to the lower flank.

DISCUSSION

The mounting evidence gathered during the past decade has emphasized the role of ocean-bottom currents on the sediment distribution on and around seamounts and topographic highs (Lowrie and Heezen, 1967; Hinz, 1969; Karig et al., 1970; Heezen and Hollister, 1971; Lonsdale et al., 1972; McGregor et al., 1973). While currents doubtlessly are important in redistributing pelagic sediment around positive topographic features, other transport mechanisms responsible for the dispersal of sediment may also be operative.

Sediment on the steeper upper- to mid-flank sectors of the Gilliss Seamount consists of volcanic rubble, organic debris and pelagic material. This sector, when observed in detail, reveals a complex, step-like morphologic character. Steep scarps and ledges (sometimes near-vertical) ranging from a few meters to several tens of meters in relief present a terrace-like profile. This step-like topography of the Gilliss Seamount was also observed during a Woods Hole Oceanographic Institution deep submersible (DSRV "Alvin") study of the New England seamounts (Heirtzler et al., 1974). The Gilliss dive, located due west of the southern peak and the upper photographic track, was restricted to a vertical distance of some 200 m (between depths of 2690 to 2487 m). The terraces result from primary volcanic processes (cf., Lonsdale et al., 1972); no evidence is found in support of a wave planation origin for these ledges of the type recorded on other shallower seamounts (Pratt, 1963; Palmer, 1964; Budinger, 1967; Schwartz, 1972; and Schwartz and Lingbloom, 1973).

In this upper sector we propose that a combination of processes, involving current traction and gravity, causes a downslope displacement of sediment. From the sum of these observations the following sediment transport scheme is outlined: sediment on the generally narrow and near-horizontal ledges is driven by bottom currents, as indicated by structures such as ripples (Fig.5C, arrow). Eventually this traction mode results in spill-over of material off ledges (Fig.5A, D, arrow, *L* = ledge) which then is transferred rapidly by gravity down steeply dipping (Fig.3A) to vertical (Fig.5B) scarps. This sediment comes to rest at the base of the scarp (Fig.5B) where it is once again entrained as part of the traction carpet on the bottom currents. This sequence of events is repeated until the sediment is finally emplaced at the base of the seamount. During this process, sediment tends to be displaced in the general direction of predominant current flow. Our model of gravity-current sediment transport is presented schematically in Fig.6.

It is useful to call attention to similar observations recorded on other seamounts where this model may be applicable. That sediments on the flank of topographic highs are moved by currents has been indicated by sedimen-

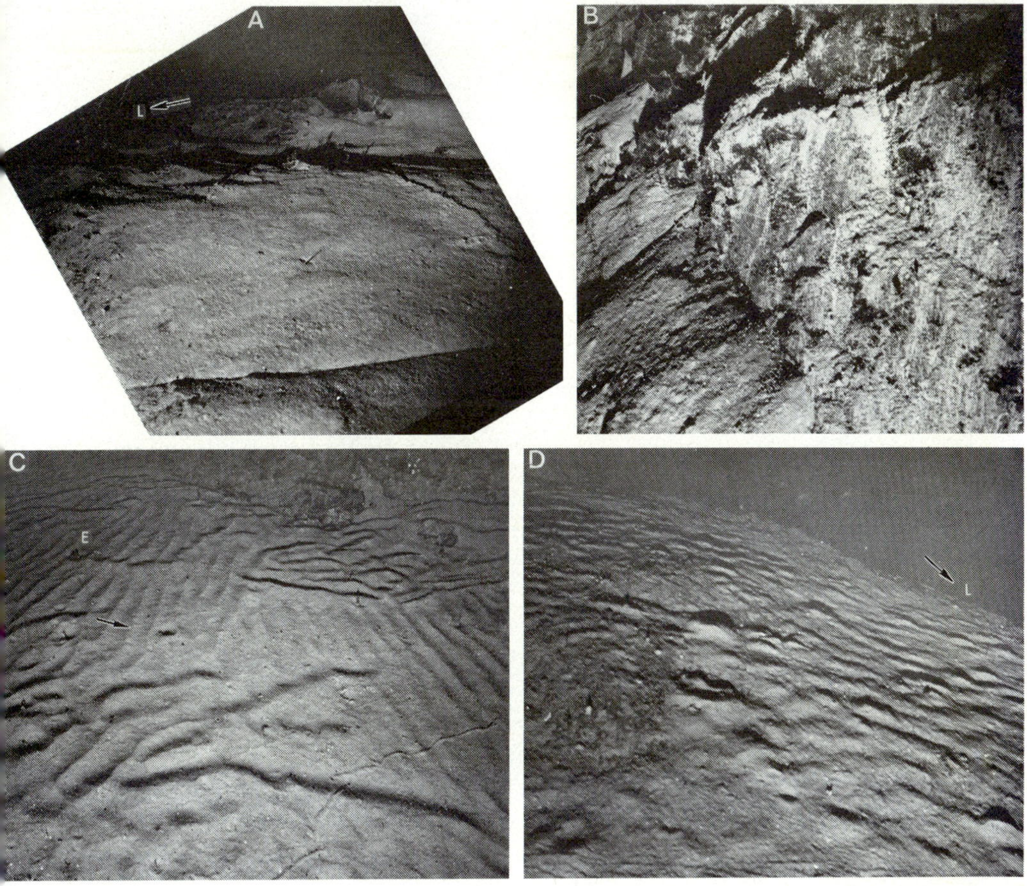

Fig.5. Sequence of photographs showing influence of gravity and currents on sediment moving on upper flank of Gilliss Seamount. Explanation in text. Note meandering tracks currently being produced by echinoid (*E* in photo C).

tary features such as ripple marks (Pratt, 1967; Heezen and Hollister, 1971; Lonsdale et al., 1972) and by areas of sediment-free igneous pillows (Uchupi, 1968; Herzer, 1971; Dangeard et al., 1973; and Schwartz and Lingbloom, 1973). Furthermore, the down-flank displacement of sediment debris under the influence of gravity can also be interpreted from the bottom photographs collected on different seamounts (Pratt, 1963; 1967; Budinger, 1967; Dangeard and Giresse, 1968; Heezen and Hollister, 1971). Descriptions by most of the above-cited authors indicate that the downslope-moving material is generally a mixture of volcanic rubble and pelagic sediments not unlike the type found on Gilliss Seamount. In many cases, this material appears to move as anastomosing tongues down scarp faces and in crevasses by rock fall (Pratt, 1967; Dangeard and Giresse, 1968) and by other forms of gravity sediment flow mechanisms. This progressive downslope process can explain

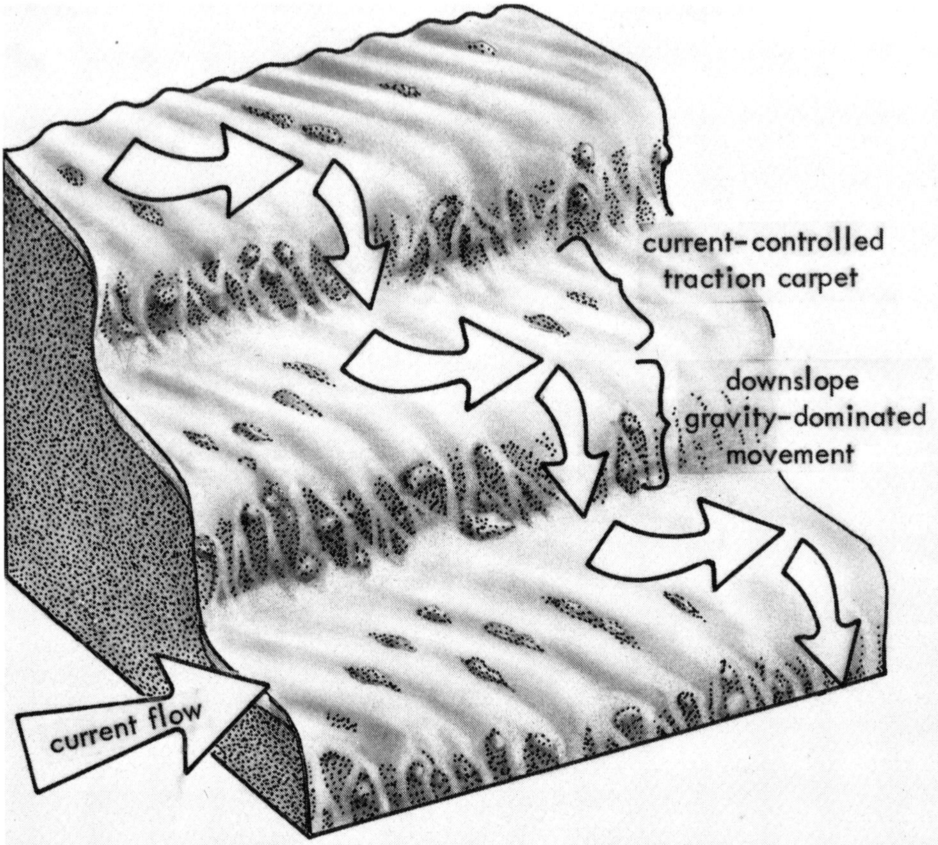

Fig.6. Schematic diagram of proposed current traction—gravity transport model of down flank sediment movement on a positive topographic feature.

the lack of significant sediment accumulations on the upper flanks of seamounts as revealed by seismic profiling (Uchupi, 1968; Hinz, 1969; Karig et al., 1970; and others).

The material in the process of accumulating at the base of mounts is commonly modulated by a regional flow of bottom water. A similar mechanism has been described for the California continental borderland (Gorsline et al., 1968). In the case of the Gilliss Seamount our photographic survey shows the influence of current transport as shown by sedimentary structures and the reduced role of gravity on the lower slopes. The importance of long-term current activity in displacing sediment is demonstrated by the asymmetric piling of acoustically transparent pelagic sediment around the base of the mount (Fig.1), as revealed in the seismic survey (Taylor et al., 1975). Bottom-current activity resulting in significant sediment accumulation also has been shown for other seamounts (Hinz, 1969; Uchupi et al., 1970; Heezen and Hollister, 1971).

From a re-evaluation of the Gilliss photographs we have tried to emphasize that the role of gravity in the movement of sediment on seamounts is a significant factor.

ACKNOWLEDGEMENTS

All photographs used in this report were taken and processed by Walter Jahn (December 1973 cruise of USNS "Lynch", U.S. Naval Oceanographic Office). Drs. J.W. Pierce (Smithsonian Institution) and W.F. Ruddiman (U.S. Naval Oceanographic Office) kindly reviewed this paper. Financial support for this project was provided by the Smithsonian Research Foundation Grant 460132 (DJS), and the Office of Naval Research (PTT).

REFERENCES

Budinger, T.F., 1967. Cobb Seamount. Deep-Sea Res., 14: 191—201.

Cordell, L. and Taylor, P.T., 1971. Investigation of magnetization and density of a North Atlantic seamount using Poisson's theorem. Geophysics, 36: 919—937.

Dangeard, L. and Giresse, P., 1968. Enseignements géologiques des photographies sous-marines. Bull. Bur. Rech. Géol. Min., Sect. 4(2): 1—85.

Dangeard, L., Deniaux, B. and Johnson, G.L., 1973. Les enseignements géologiques de la photographie sous-marine: le volcanisme sous-marin. Ann. Inst. Océanogr., 49: 77—82.

Dott Jr., R.H., 1963. Dynamics of subaqueous gravity depositional processes. Am., Assoc. Pet. Geol. Bull., 47: 104—128.

Gorsline, D.S., Drake, D.E. and Barnes, P.W., 1968. Holocene sedimentation in Tanner Basin, California continental borderland. Geol. Soc. Am. Bull., 79: 659—674.

Grigg, R.W., 1972. Orientation and growth form of sea fans. Limnol. Oceanogr., 17: 185—192.

Heezen, B.C. and Hollister, C.D., 1971. The Face of the Deep. Oxford University Press, London, 659 pp.

Heirtzler, J.R., Ballard, R.D., Houghton, R.L. and Taylor, P.T., 1974. An investigation of the New England seamounts by submersible. Trans. Am. Geophys. Union, 56: 1138 (abstract).

Herzer, R.H., 1971. Bowie Seamount: a recently active flat-topped seamount in the Northeast Pacific Ocean. Can. J. Earth Sci., 8: 676—687.

Hinz, K., 1969. The Great Meteor Seamount. "Meteor" Forschungsergeb., Reihe C, 2: 63—77.

Karig, D.E., Peterson, M.N.A. and Shor, G.G., 1970. Sediment-capped guyots in the Mid-Pacific Mountains. Deep-Sea Res., 17: 373—378.

Lonsdale, P., Normark, W.R. and Newman, W.A., 1972. Sedimentation and erosion on Horizon Guyot. Geol. Soc. Am., Bull., 83: 289—316.

Lowrie Jr., A. and Heezen, B.C., 1967. Knoll and sediment drift near Hudson Canyon. Science. 157: 1552—1553.

McGregor, B.A., Betzer, P.R. and Krause, D.C., 1973. Sediment in the Atlantic Corner seamounts: control by topography, paleo-winds, and geochemically-detected modern bottom currents. Mar. Geol., 14: 179—190.

Middleton, G.V. and Hampton, M.A., 1973. Sediment gravity flows: mechanics of flow and deposition. In: G.V. Middleton and A.H. Bouma (Editors), Turbidites and Deep-Water Sedimentation. Pacific Section, S.E.P.M., Los Angeles, Calif., pp.1—38.

Palmer, H.D., 1964. Marine geology of Rodriguez Seamount. Deep-Sea Res., 11: 737—756.

Pratt, R.M., 1963. Great Meteor Seamount. Deep-Sea Res., 10: 17—25.

Pratt, R.M., 1967. Photography of seamounts. In: J.B. Hersey (Editor), Deep-Sea Photography. Johns Hopkins Press, Baltimore, Md., pp.145—158.

Schwartz, M.L., 1972. Seamounts as sea-level indicators. Geol. Soc. Am. Bull., 83: 2975—2980.
Schwartz, M.L. and Lingbloom, K.L., 1973. Research submersible reconnaissance of Cobb Seamount. Geology, 1: 31—32.
Stanley, D.J., Fenner, P. and Kelling, G., 1972. Currents and sediment transport at the Wilmington Canyon shelfbreak, as observed by underwater television. In: D.J.P. Swift, D.B. Duane and O.H. Pilkey (Editors), Shelf Sediment Transport: Process and Pattern. Dowden, Hutchinson and Ross, Stroudsburg, Pa., pp.621—644.
Taylor, P.T. and Hekinian, R., 1971. Geology of a newly discovered seamount in the New England seamount chain. Earth Planet. Sci. Lett., 11: 73—82.
Taylor, P.T., Stanley, D.J., Simkin, T. and Jahn, W., 1975. Gilliss Seamount: detailed bathymetry and modification by bottom currents. Mar. Geol., 19: 139—157.
Uchupi, E., 1968. Long lost Mytilus. Oceanus, 14: 1—7.
Uchupi, E., Phillips, J.D. and Prada, K.E., 1970. Origin and structures of the New England seamount chain. Deep-Sea Res., 17: 483—494.

Marine Geology, 23 (1977) 89—101

CIRCUM-POLAR CIRCULATION AND LATE TERTIARY CHANGES IN THE CARBONATE COMPENSATION DEPTH SOUTH OF AUSTRALIA*

CLIFFORD D. MALLET and BRUCE C. HEEZEN

University of Melbourne, Melbourne, Vic. (Australia)
Lamont-Doherty Geological Observatory of Columbia University, Palisades, N.Y. (U.S.A.)

(Received April 28, 1976)

ABSTRACT

Mallet, C.D. and Heezen, B.C., 1977. Circum-polar circulation and late Tertiary changes in the carbonate compensation depth south of Australia. Mar. Geol., 23: 89—101.

Cores in the southern Indian Ocean south of Australia show that the carbonate compensation depth lies at 4800 m, controlling the distribution of pelagic ooze and red clay. Above the compensation depth, the effect of differential solution on foraminiferal tests is reflected in the core-top assemblages obtained at various depths. At some sites reworking of the faunas makes it difficult to apply the solution susceptibility series as an index of carbonate solution. Below 4000 m Quaternary oozes overlie unfossiliferous clay. This suggests that the present compensation depth lies approximately 800 m below the position occupied in the Pliocene. Possible controls of the compensation depth in the area are evaluated, and it is concluded that variations in the circum-polar water circulation in the Quaternary is the dominant factor.

CORE LOCATIONS

Piston cores were obtained between latitudes 32°S—40°S and longitudes 110°E—140°E during USNS "Eltanin" cruise 55 (Fig.1). The cores were sited on the continental slope, the Great Bight Abyssal Plain, the Diamantina Fracture Zone, and the flanks of the Naturaliste Plateau. Depth of water at core sites ranged from 3400 m to 5300 m. In an effort to penetrate through Quaternary cover, sites were chosen on scarps or flanks of seamounts, where seismic profiles indicated a thinning of sediments. Several cores were taken on each seamount studied.

LITHOLOGIES

The pelagic sediments were logged as foraminiferal and nannofossil ooze, carbonate-bearing clays, or carbonate-free red clay.

The oozes were white or light tan in colour. The texture varied with the nannofossil content, and the stiff white nannofossil-rich oozes occurred at shallower depths than foraminiferal oozes. The light-tan oozes were composed primarily of foraminiferal tests, with the addition of a small percentage of clay.

---

*Contribution No. 2373 of the Lamont-Doherty Geological Observatory.

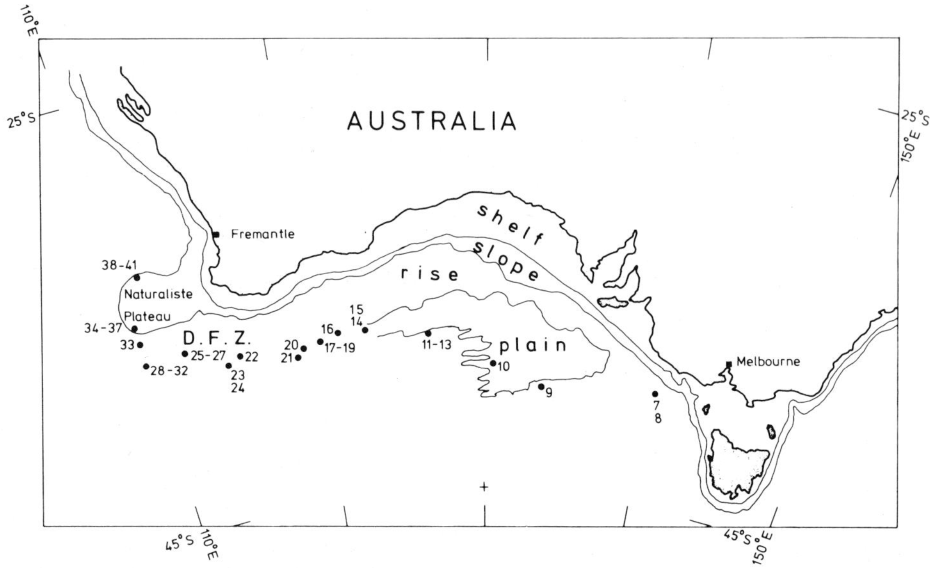

Fig.1. Location of "Eltanin" cruise 55 core sites. (*D.F.Z.* = Diamantina Fracture Zone.)

Clays ranged in colour from light to dark brown, with the darkest material stiff and consolidated. No carbonate is preserved in these sediments, which must have been deposited below the carbonate compensation depth. The sediment consisted mainly of brown clay, with minor amounts of manganese micronodules, fish teeth, wind-blown quartz, and the siliceous radiolaria and diatoms.

Intermediate between these lithologies was yellow to brown calcareous clay. The carbonate tests within the clay usually showed marked solution effects, and a reduction in the number of species preserved. The lithology grades on the one hand into ooze and on the other into red clay. The boundary with ooze is arbitrary, but samples included in the intermediate category would usually consist of more than 50% clay. A sample was logged as red clay if no foraminifera or carbonate was present in the +50 μm fraction.

The distribution of the lithologies is given in Fig.2.

BIOSTRATIGRAPHY

The only foraminiferal datum recognised in the cores was the initial appearance of *Globorotalia truncatulinoides* (d'Orbigny), which is considered as representing the base of the Pleistocene. In southern Australia, Pliocene foraminiferal faunas are characterised by strongly carinate descendants of *Globorotalia crassiformis* (Galloway and Wissler), (*Globorotalia crassula viola* Blow, and inflated forms similar to *Globorotalia conomiozea* Kennett, with thicker imperforate carinae). These forms become rare after the appearance of *G. truncatulinoides.* The late Pliocene index foraminifera *Globorotalia*

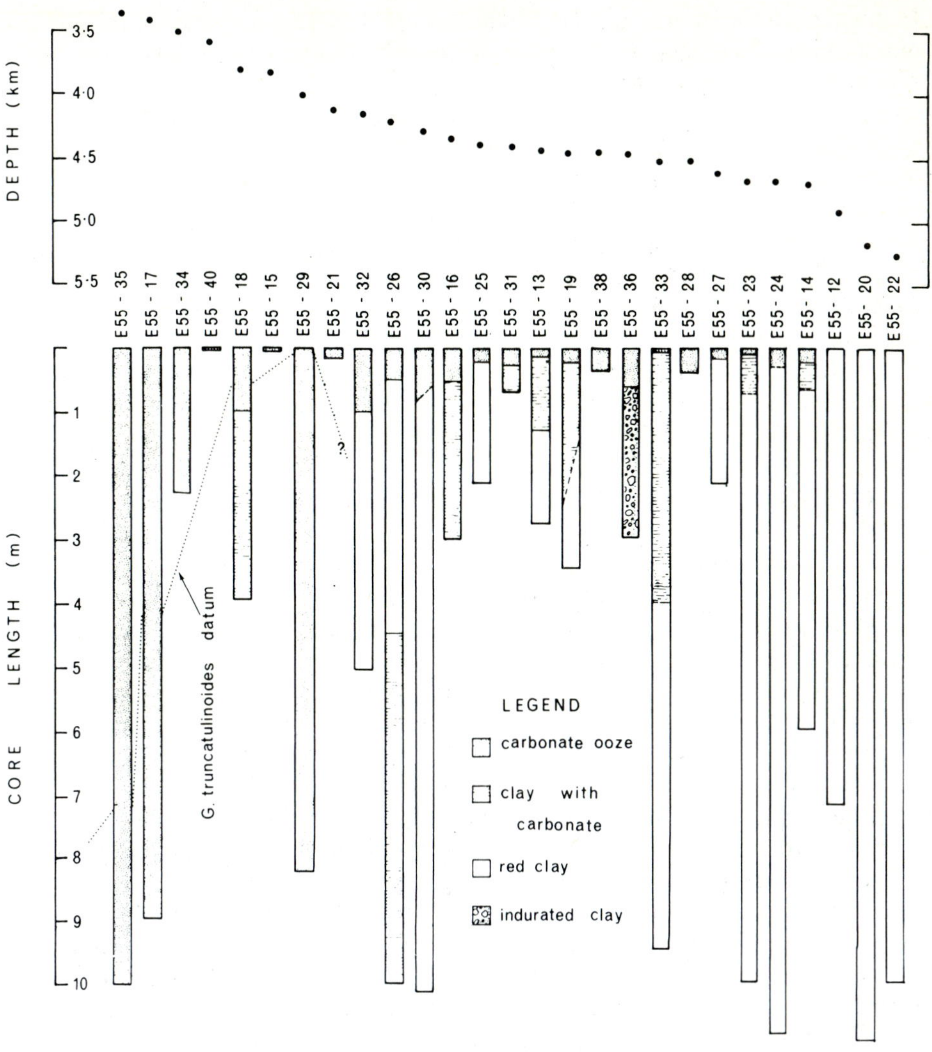

Fig.2. Summary of "Eltanin" 55 core logs.

tosaensis Takayanagi and Saito, has not been found in southeastern Australia, but occurred with or just below the first appearance of *G. truncatulinoides* in three cores, E55-17, E55-29, E55-35, all west of 120° E.

Hays and Berggren (1971) summarised the evidence for using the initial appearance of *G. truncatulinoides* to identify the base of the Pleistocene. Although they accept the first appearance as a valid criterion in the tropics, they suggest that it is only of moderate to poor value in the latitude of these samples. Fossiliferous sections interbedded with radiometrically dated lavas in southern Australia have been described. The sediments older than

2.4 m.y. contained no *G. tosaensis* or *G. truncatulinoides* despite a rich planktonic assemblage being present, whereas 2.2 m.y. contained *G. truncatulinoides* but no *G. tosaensis*. This is consistent with an initial appearance of 1.8 m.y. for *G. truncatulinoides* in the area, as has been reported from tropical areas.

Kennett and Geitzenauer (1969) and Theyer (1973) have reported *G. truncatulinoides* occurring within the Pliocene in the South Pacific Ocean. Theyer suggests that *G. tosaensis* and *G. truncatulinoides* form a plexus representing keeled and non-keeled forms. The plexus appeared approximately 3.0—3.5 m.y. ago, represented by *G. tosaensis* in low latitudes and both *G. tosaensis* and *G. truncatulinoides* in higher latitudes. It was not until the base of the Pleistocene that *G. truncatulinoides* migrated into tropical waters.

Watkins et al. (1973) have questioned Theyer's evidence, and maintain that *G. truncatulinoides* does represent a consistent index for the base of the Pleistocene in the South Pacific.

CALCIUM CARBONATE SOLUTION INDICES

Murray and Renard (1891) noted a depth control of the solution of carbonate in the oceans, and recognised a carbonate compensation depth, based on the percentage of carbonate remaining in a sediment. Although factors other than depth influence the percentage of carbonate in a sediment, the compensation depth, or critical depth, has proved a useful parameter in mapping ocean-floor sediments.

Attempts have been made to accurately establish the solution of carbonate in oceanic waters, using the differing solution rates of different carbonate particles. Some of the criteria used have been: the number of whole foraminiferal specimens in unit weight of sediment, damage to foraminiferal tests, and the percentages of foraminiferal species which are controlled by selective solution (Berger, 1968; Olausson, 1971). Experiments in the Pacific Ocean were reported by Peterson (1966) and Berger (1967), when calcite spheres and sediment samples were lowered to various depths and the solution effects noted. Berger (1968) described the lysocline, which is the depth at which the maximum effect of solution on the foraminiferal assemblage is noticed.

The assemblages of foraminifera in core-top samples were investigated and the percentages of species in counts of 300 were obtained. The taxonomy used follows that of Parker (1962). The solution indices of the assemblages were calculated following the method outlined by Berger (1968), and plotted against depth from which they were obtained (Fig.3) (where the position of a species in the solution susceptibility series was not known, it was given the mean susceptibility value of the whole sample to avoid biasing the solution index). The points do not lie on a simple curve which indicates the lysocline, as might be expected from Berger's results. Points are sufficiently close and consistent below 4000 m, however, to indicate the lysocline is probably between 4300 and 4400 m.

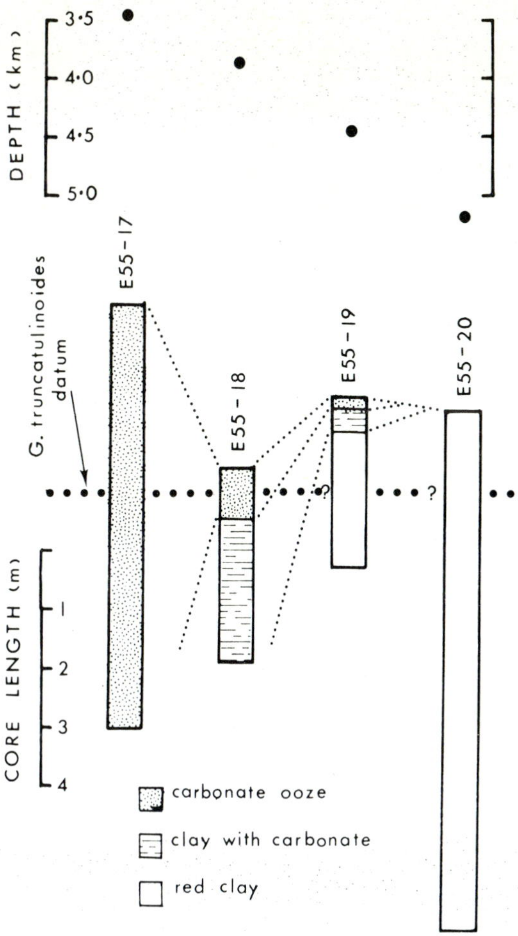

Fig.3. Summary of cores E55-17—19 taken on Molar Seamount near 27°30′S—120°10′E and core E55-20 taken in the adjacent abyssal hills south of the Great Bight Abyssal Plains at 37°50′S—119°00′E.

There is considerably more variation in the solution index values than would be expected from the results of Berger (op. cit.). Errors in the solution susceptibility series or variations in the primary composition of the fauna are not sufficient to cause the large deviations from expected values obtained in some samples. A possible explanation is provided by core-top sample E55-29. Anomolously high values of the solution index were produced in this sample by reworking of the sediment during recovery. Apart from variations in the primary fauna considered by Berger, any mechanical sorting will alter the composition of a fauna and thus the value of its solution index. The effect of sorting on the solution index will depend on the initial composition of the fauna. The living faunas from which these sediments were derived fall into the transitional-zone assemblages of Bé and Tolderlund (1971). In these, mechanical sorting leads to concentration of the larger specimens of *G. truncatulinoides* and *G. inflata*, which raises the solution index. All the samples were obtained from slopes and scarps and so are likely to be subject to mechanical reworking by currents or slumping, and this could account for high values for the solution indices above the lysocline.

The distribution of common species in core-top samples shows (Fig.4) that there is no simple relationship to depth. In general, the percentages of *G. inflata* and *G. truncatulinoides* increase with depth, while most other species decrease. Marked changes in assemblages occur at two depths. At 3500 m most species show a small drop in percentage, while *G. inflata*, *G. truncatulinoides* and *Globigerina bulloides* d'Orbigny 1826, show a small increase. However, *Globigerina quinqueloba* Natland 1938 shows a large increase, apparently not as a result of selective solution as other resistant species do not show corresponding increase. Specimens of *G. quinqueloba* are usually smaller than other species, but it is unlikely that they are winnowed from the shallower samples, which are mainly nannofossil oozes, and show no obvious effects of reworking. Another puzzling aspect of the high percen-

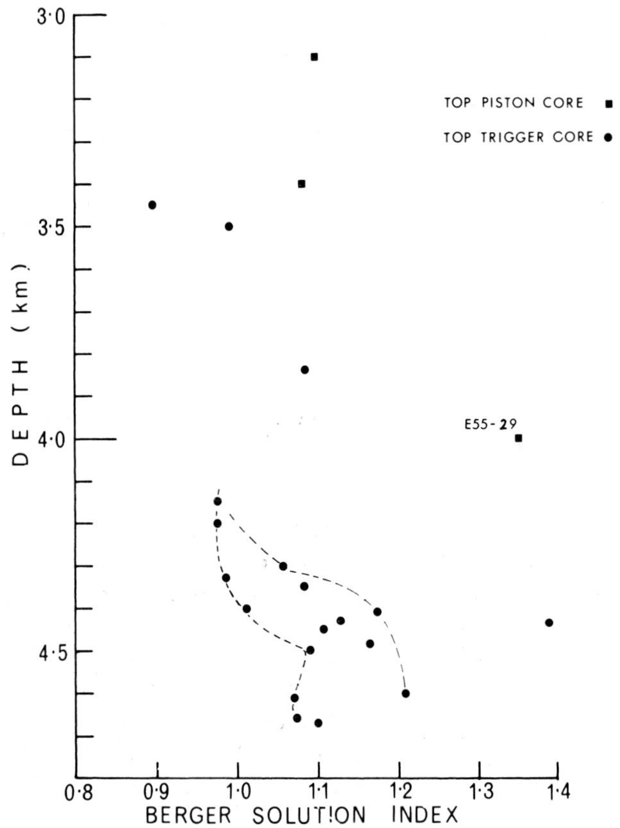

Fig.4. Plot of solution index (Berger, 1968) against depth for core-top samples.

tages of *G. quinqueloba* is that the species is at present extremely rare in planktonic tows in the latitude of the cores. In present-day surface waters *G. quinqueloba* occurs predominantly at temperatures colder than 12°C, and in the Antarctic region, the greatest frequency is encountered between 1°C and 5°C (Bé and Tolderlund, 1971). As the surface temperature in the sampled area ranges from 12°C to 19°C (Netherlands Met. Inst., 1947), it appears unlikely that the tests of *G. quinqueloba* in the sediment are being produced in the overlying surface waters. The small specimens are easily transported by currents, and it may be that they are brought into the area by deep currents from higher latitudes. The salinity maximum taken to indicate the axis of the Antarctic circum-polar current lies at 3000 m south of Western Australia (Gordon, 1972b) and this represents the most likely medium for transporting the *G. quinqueloba*.

The second change in the fauna occurs in the depth range 4400—4500 m. Here there is an increase in the percentage of *G. inflata* (to 60—70%), and a decrease in the percentages of *G. quinqueloba*, *G. bulloides* and *Globigerinita glutinata* (Egger) 1893. The distribution of species in these samples is similar to that in the core top of piston core E55-29, which was washed during recovery. The samples do not reflect the trends in samples above and below this depth. It is considered that the samples in this interval have been winnowed, concentrating the large *G. inflata*.

VARIATION OF SEDIMENT TYPE WITH DEPTH AND TIME

At present the carbonate compensation depth lies at approximately 4800 m, with carbonate oozes forming above this depth, and unfossiliferous clays below. Above 3500 m the dominant component of oozes is nannofossil remains, but their content decreases with increasing depth. Oozes in the range 4500—4800 m are typically buff and tan coloured, as clay is mixed with the carbonate. In many cores these coloured oozes are only thin veneers containing *G. truncatulinoides* on unfossiliferous clays which indicate a change in sediment type during the Pleistocene and Recent. These changes are illustrated in Fig.2.

Four depth-controlled lithology groups are recognised.

(1) Depths less than 4000 m. Present sedimentation at these depths is carbonate ooze. This type of deposition has occurred throughout the Pleistocene and Pliocene where it has been penetrated. One core at these depths showed carbonate-bearing clay in the Pliocene (E55-18).

(2) Depths 4000—4500 m. Carbonate ooze is being formed in this depth range; 20—50 cm of ooze overlies 1—4 m of calcareous clay, which in turn overlies brown clay. All fossiliferous samples are above the *G. truncatulinoides* datum.

(3) Depths 4500—4800 m. Present sedimentation is tan carbonate ooze, and it overlies a thin layer of calcareous clay. The fossiliferous section is less than 0.5 m thick, and post-dates the *G. truncatulinoides* datum.

(4) Depths greater than 4800 m. All cores are composed of red clay.

This facies change is illustrated in Fig.5, which summarises four adjacent cores, E55-17, 18, 19, 20. The ooze and carbonate-bearing clay has accumulated at successively greater depths during the Plio-Pleistocene. The present compensation depth is approximately 800 m lower than its position at the *G. truncatulinoides* datum. The depth of accumulation of calcareous clay has moved in a similar manner.

Schott (1939, his fig.1) shows globigerina ooze overlying red clay on swales of the mid-oceanic ridge, southeast of Madagascar. This is at the boundary of areas of red clay and ooze, and represents encroachment of carbonate into deeper areas of previous complete carbonate solution.

In the South Pacific Ocean, Koster (1966) reported several cores in which carbonate ooze overlies clay or siliceous ooze. In some cases ooze is interbedded with clays in the late Pleistocene and Recent. These cores are in the vicinity of the Antarctic Convergence, and the sediments can be related to migration of the convergence with associated changes in the supply of carbonate. Fisher (1968) showed that in the South Pacific Ocean the depth of carbonate deposition is now greater than it was at the time of the boundary between the Brunhes and Matuyama magnetic epochs.

Berger (1968) suggested that there has been a depression of approximately 400 m in the lysocline in the Atlantic, since the last glacial stage. This is based on one core from approximately 4000 m, in which the fauna below 30 cm shows considerably greater solution effects than the core top.

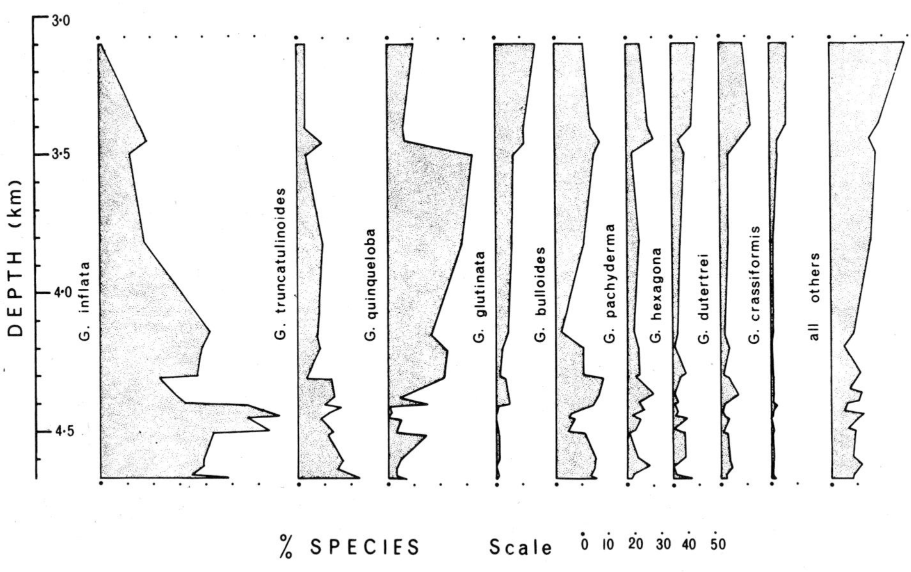

Fig.5. Distribution of species in core-top samples related to depth of core.

CARBONATE COMPENSATION DEPTH CONTROLS

Primary productivity

The deposition of carbonate at depth must depend to some extent upon the amount of carbonate produced at the surface of the oceans. It might be expected that an increase in primary production could depress the compensation depth. The area off southwestern Australia is intermediate between the highly productive zone around the Antarctic Convergence, and the barren areas in the tropical Indian Ocean. Seasonal variations in productivity have been reported (CSIRO 1963, 1965a, b, c, 1966a, b, c, 1967), and are illustrated in Fig.6 and 7. Extremely low productivity is associated with the strong westerly drift in winter and spring, but this increases with the greater north—south water movement that develops in summer. The presently fairly low production in the area does not support the suggestion that the depression of the compensation depth is caused by an increased supply of carbonate. Variation of carbonate in cores has often been attributed to the effects of changing climate in the Pleistocene glacial fluctuations (Arrhenius, 1963; Olausson, 1971). The climatic fluctuations would affect oceanic circulation, which would in turn change the pattern of carbonate production in the oceans. To achieve large increases in the supply of organic carbonate, the highly productive zones associated with the Antarctic Convergence would have to migrate a considerable distance north. Berger (1970) argued that increasing productivity would not necessarily increase the deposition of carbonate, as the associated increase in organic matter supplied to the sea floor would aid the solution of carbonate due to the increased benthonic activity. There is little evidence to suggest that the primary production of carbonate has controlled the compensation depth.

Carbonate budget

Olausson (1971) considered the problem in terms of the overall calcium carbonate budget of the worlds oceans. He postulated that in glacial times, carbonate is dissolved from the North Atlantic Ocean and added to the available supply in the southern hemisphere of the other oceans, increasing the deposition of carbonate there. The process is reversed during interglacials. He described cores illustrating these changes. In this way changes in the primary production of calcium carbonate occur independently of the variations in the overall primary production in an area. If this alternation in production of carbonate has operated south of Australia, a thin veneer of ooze deposited during a glacial stage may be dissolved in the following interglacial, due to fluctuations in the compensation depth. Blanketing by residual clays, and insufficient exposure time could give incomplete solution, forming calcareous clays. Detailed sampling of cores would show if such alternations were present.

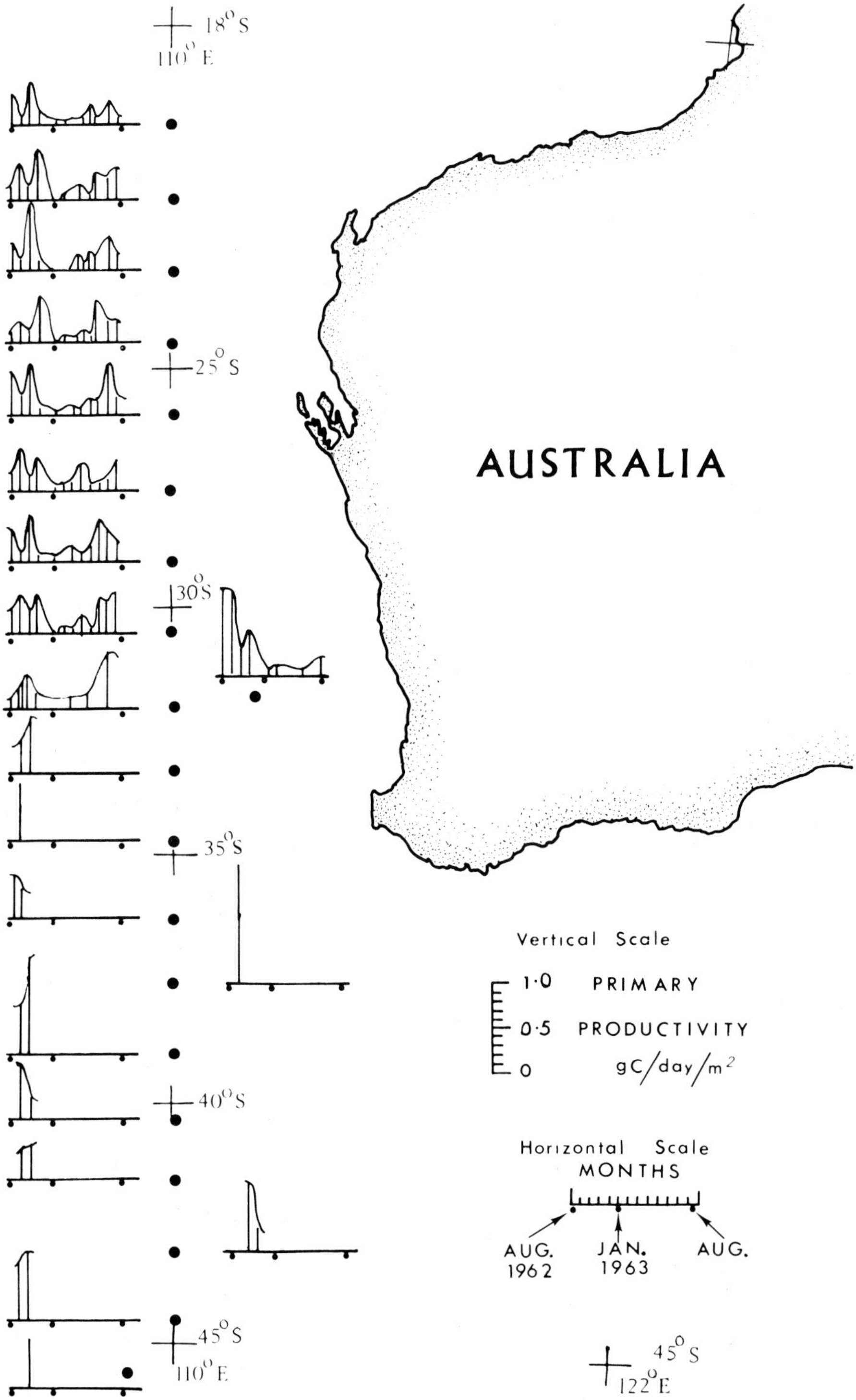

Fig.6. Primary production off Western Australia as determined by C.S.I.R.O. (1965a, b, c, 1966a, b, 1967) using the ^{14}C method (Dyson et al., 1965). Samples were exposed to four hours artificial light, and values calculated as grams of carbon per day per square meter, to a depth of 150 m, based on six samples at various depths.

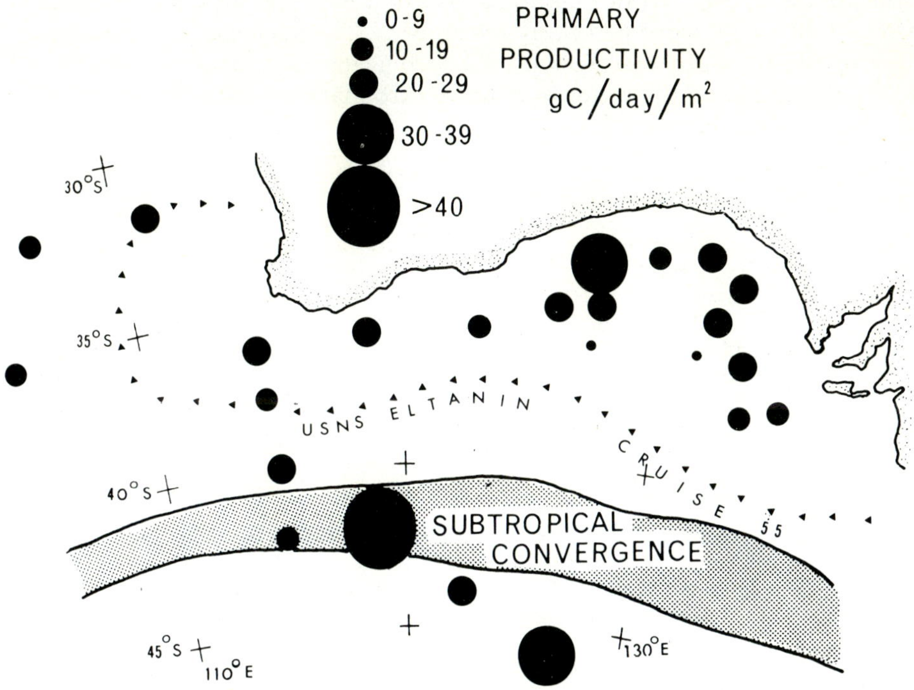

Fig.7. Primary production south of Western Australia as determined by C.S.I.R.O. (1963, 1966c) using the ^{14}C method (Dyson et al., 1965). Samples were exposed to four hours artificial light and values calculated as grams of carbon per day per square meter to a depth of 150 m, based on six samples at various depths. The position of the subtropical convergence is after Gordon (1972).

Currents and water masses

Local changes in currents and the distribution of water masses could alter the ability of bottom water to dissolve calcium carbonate, giving rises and falls in the compensation depth. Erosion or non-deposition caused by high current activity may have been intermittent, allowing alternating solution and deposition at the depths studied. This would account for the thin condensed sections of Quaternary ooze obtained.

G. tosaensis occurs with or just below the first appearance of *G. truncatulinoides* in cores E55-17, 29, 35. The restricted occurrence of *G. tosaensis* could be due to its time range corresponding to a phase of non-deposition, rather than it being ecologically excluded from the area. Watkins and Kennett (1972) suggested that high circum-polar current velocities have produced a regional unconformity on the mid-ocean ridge south of Tasmania, exposing Pliocene sediments. Bottom photographs of areas affected by erosion show high current velocity features, but where there was a thin cover of Bruhnes magnetic epoch sediments, only low velocity features were observed. It seems that the current now operating has lower energy than that operating during the Late Pliocene and Early Pleistocene. The decrease in resident time for bottom waters associated with such a current weakening, would reduce the ability of the water to dissolve carbonate, and thus cause the lowering of the compensation depth.

REFERENCES

Arrhenius, G., 1963. Pelagic sediments. In: M.N. Hill (Editor), The Sea, 3. Interscience, New York, N.Y., pp.655—718.

Bé, A.W.H. and Tolderlund, D.S., 1971. Distribution and ecology of living planktonic foraminifera in surface waters of the Atlantic and Indian Oceans. In: B.M. Funnell and W.R. Riedel (Editors), The Micropalaeontology of the Oceans. Cambridge Univ. Press, London, pp. 105—149.

Berger, W.H., 1967. Foraminiferal ooze: solution at depths. Science, 156: 383—385.

Berger, W.H., 1968. Planktonic foraminifera: selective solution and palaeoclimatic interpretation. Deep-Sea Res., 15: 31—43.

Berger, W.H., 1970. Planktonic foraminifera: selective solution and the lysocline. Mar. Geol. 8: 111—138.

C.S.I.R.O. Aust. Oceanogr. Cruise. Reports, 1963, No. 7; 1965a, No. 23; 1965b, No. 24; 1965c, No. 25; 1966a, No. 17; 1966b, No. 18; 1966c, No. 10; 1967, No. 20.

Dyson, N., Jitts, H.R. and Scott, B.D., 1965. Techniques for measuring oceanic primary production using radioactive carbon. C.S.I.R.O. Aust. Div. Fish. Oceanogr. Tech. Pap. No. 18.

Fisher, V.A., 1968. The Southern Ocean 700,000 Years Ago. Thesis, Univ. of Florida, Dept. of Geology, Sedimentological Laboratory Contribution No. 28.

Funnell, B.M. and Riedel, W.R., 1971. The Micropalaeontology of the Oceans. Cambridge Univ. Press, London, 823 pp.

Gordon, A.L., 1972a. Physical oceanography of the Southeast Indian Ocean. In: Antarctic Oceanology, 2. The Australian—New Zealand Sector. Antarctic Res. Ser., 19: 3—9.

Gordon, A.L., 1972b. On the interaction of the Antarctic Circumpolar Current and the Macquarie Ridge. In: Antarctic Oceanology, 2. The Australian—New Zealand Sector. Antarctic Res. Ser., 19: 3—9. op. cit. pp. 71—78.

Hays, J.D. and Berggren, W.A., 1971. Quaternary boundaries and correlations. In: B.M. Funnell and W.R. Riedel (Editors), The Micropalaeontology of the Oceans. Cambridge Univ. Press, London, pp.669—691.

Hill, M.N., 1963. The Sea. Interscience, New York, N.Y., 4 vols.

Kennett, J.P. and Geitzenauer, K.R., 1969. Pliocene—Pleistocene boundary in a South Pacific deep-sea core. Nature, 224: 899—901.

Koster, S., 1966. Recent Sediments and Sedimentary History Across the Pacific—Antarctic Ridge. Thesis, Univ. of Florida, Dept. of Geology, Sedimentary research laboratory, research contribution No. 17.

Lisitizin, A.P., 1971. Distribution of carbonate microfossils in suspension and bottom sediments. In: B.M. Funnell and W.R. Riedel (Editors), The Micropalaeontology of the Oceans. Cambridge Univ. Press, London, pp.197—218.

Murray, J. and Renard, A.F., 1891. Deep sea deposits. Sci. Rep. Challenger Exped., 1873—1876.

Netherlands Meteorological Institute, 1947. Sea Areas Round Australia. Publ. No. 124.

Olausson, E., 1971. Quaternary correlations and the geochemistry of oozes. In: B.M. Funnell and W.R. Riedel (Editors), The Micropolaeontology of the Oceans. Cambridge Univ. Press, London, pp.375—398.

Parker, F.L., 1962. Planktonic foraminiferal species in Pacific sediments. Micropaleontology, 8: 219—254.

Peterson, M.N.A., 1966. Calcite: rates of dissolution in a vertical profile in the central Pacific. Science, 154: 1542—1544.

Schott, W., 1939. Deep sea sediments of the Indian Ocean. In P.D. Trask (Editor), Recent Marine Sediments. A.A.P.G., Tulsa, Okla., pp. 396—408.

Theyer, F., 1973. *Globorotalia truncatulinoides* datum plane: Evidence for a Gauss (Pliocene) age in subantarctic cores. Nature Phys. Sci., 241: 142—145.

Trask, P.D. (Editor), 1939. Recent Marine Sediments. A.A.P.G., Tulsa, Okla., pp.1—736.

Watkins, N.D. and Kennett, J.P., 1972. Regional sedimentary disconformities and Upper Cenozoic changes in bottom water velocities between Australia and Antarctica. In: Antarctic Oceanology, 2. The Australian—New Zealand Sector. Antarctic Res. Ser., 19: 273—293.

Watkins, N.D., Kennett, J.P. and Vella, P., 1973. Palaeomagnetism and the *Globorotalia truncatulinoides* datum in the Tasman Sea and Southern Ocean. Nature Phys. Sci., 244: 45—48.

Marine Geology, 23(1977) 103—111

EROSION OF DEEP-SEA SEDIMENTS IN THE SOUTHERN OCEAN BETWEEN LONGITUDES 70°E AND 190°E AND CONTRASTS IN MANGANESE NODULE DEVELOPMENT

N.D. WATKINS and J.P. KENNETT

Graduate School of Oceanography, University of Rhode Island, Kingston, 02881 R.I. (U.S.A.)

(Received July 1, 1976)

ABSTRACT

Watkins, N.D. and Kennett, J.P., 1977. Erosion of deep-sea sediments in the Southern Ocean between longitudes 70°E and 190°E and contrasts in manganese nodule development. Mar. Geol., 23: 103—111.

Using our previously published results, we present trend surface maps of surface sediment age, and the thickness of sediment deposited during the last 0.7 m.y. for the area between the Kerguelen Plateau, Broken Ridge, Australia, New Zealand, and Antarctica. Four separate centers of widespread deep-sea sediment erosion have developed in response to high bottom-water velocities. Two of these, centered on the South Tasman Basin and the western end of the South Australian Basin, respectively, are associated with extensive manganese nodule development. The manganese nodule fields are almost entirely developed on sediment which averages at least 0.5 m.y. in age. In contrast, the southern and eastern flanks of the Kerguelen Plateau, the Ross Sea and the Campbell Plateau have very limited manganese nodule deposits, despite abundant evidence of widespread active erosion. It is concluded that sediment erosion is a necessary but insufficient requirement for extensive manganese nodule development in the area, and that an important role is also played by submarine volcanism.

INTRODUCTION

In a series of earlier publications (Watkins and Kennett, 1971, 1972; Kennett and Watkins, 1975, 1976) we presented the ages of over 300 sediment cores, collected during cruises of the USNS "Eltanin" south and west of New Zealand and Australia (Fig.1) during the years 1965 to 1972. The total length of core involved is over 2000 m. We employed paleomagnetic and micropaleontological dating of the cores to demonstrate the existence of periods of non-deposition or erosion during the past few million years. These hiatuses were attributed to Antarctic Bottom Water (AABW) activity during the past 2.5 m.y., and this in turn led us to speculate that the Early and Middle Cenozoic deposits of the area would feature widespread hiatuses, developed as the result of substantial changes in Antarctic circum-polar

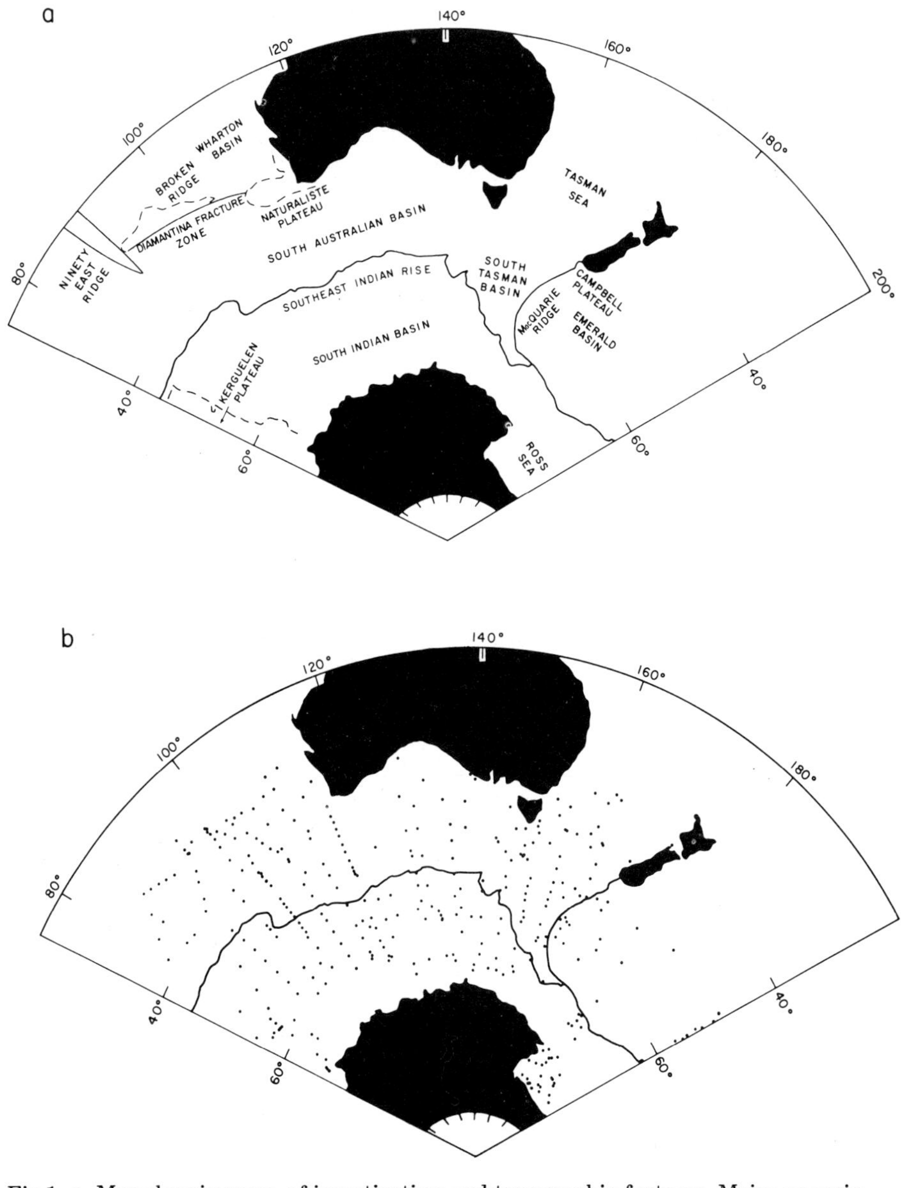

Fig.1. a. Map showing area of investigation and topographic features. Major oceanic tectonic elements indicated by solid lines: major aseismic positive elements by dashed lines. b. Map showing location of "Eltanin" cores used in this study.

circulation as Australia separated from Antarctica (Watkins and Kennett, 1972, 1973).

In this presentation, we combine all the previous results to provide a unique set of data ideally distributed for trend surface analysis (Fig.1b). The reasons for the employment of trend surface analyses are many (e.g. Watkins and Self, 1971, 1972) but the essential motive is to delineate broad features, at the expense of local perturbations not related to major controlling factors. Such local effects could include slumping, and other tectonic disturbances. We shall apply this analytical technique to the age of the sediment at the top of each core, and to the Brunhes Epoch isopach (or the thickness of sediment deposited during the past 0.7 m.y.), in an attempt to delineate the relationship between the several large-scale scour patterns which we previously defined separately in the smaller-scale studies. Regional sedimentary trends are then compared with patterns of manganese nodule distribution in the area in an attempt at understanding the genesis of the nodule fields. The extent and environmental diversity of the area involved (Fig.1) provides an opportunity to compare manganese nodule development with a wide range of possibly relevant parameters, such as water depth, surface sediment type, sedimentary history and proximity to active volcanic sea floor.

METHODS AND RESULTS

The shipboard and laboratory techniques have been described in detail in Watkins and Kennett (1972) and Kennett and Watkins (1976). The age of the top of each core was defined to about ±0.25 m.y. for cores less than 5.2 m.y. old (the base of the Gilbert Epoch) by comparison of the paleomagnetic data with the known polarity time scale, and resolution of ambiguities by micropaleontological means. The ages of older cores could not be as accurately defined, but for reasons given below, all cores older than 5.2 m.y. have been excluded from analysis. The thickness of sediment deposited during the Brunhes Epoch (t = 0—0.7 m.y.) was readily determined from the paleomagnetic data in almost all cases. For those cores with youngest sediment between 5.2 and 0.7 m.y. age, however, a negative sediment thickness was computed, using the known age, and an assumed sedimentation rate of 0.5 cm per 1000 years for the missing interval. When more than one hiatus is present in a core, the youngest is employed. Young veneer deposits, reaching usually only a few cm in thickness, were not considered.

Data from cores of two types were not included in the analysis. Those which are of Brunhes age throughout and which are less than 5 m in length were rejected because of the possibility that inferred sedimentation rates might be too small by a factor of two or more. The base of cores confined to the Brunhes Epoch, but which are longer than 5 m, were arbitrarily assigned an age of 0.6 m.y. Cores with sediment tops older than 5.2 m.y. were also rejected from the analysis not only because of the difficulty in obtaining accurate ages, but also because of the distorting effect which very old cores can have on trend surface analyses: for example, a few cores of Cretaceous age would, if included in the analysis, dominate the surface configuration to an extent which could greatly distort any trend reflecting the effect of erosion of sediments of Pliocene and Pleistocene age.

The exact location, length, and water depth for each core as well as discussion of the effect of the data selectivity on trend surface configuration are given in Watkins and Kennett (1972) and Kennett and Watkins (1976). All paleomagnetic data, and assigned age ranges for all cores are given in diagrammatic form in the same publications. In Figs.2a, 3a we present the the fifth order trend surface of the sediment age in m.y. at the top of each core, and the Brunhes isopach, in meters, respectively. The selection of the order of the surface is by conventional means: the root mean square of the residual is computed for successively higher orders, until the residual is not significantly decreased by a higher order fit. In Figs.2b, 3b we reproduce the trend surface maps but superimpose the distribution of manganese nodules, as determined from sea floor photography (Goodell, 1973; Kennett and Watkins, 1976).

DISCUSSION AND CONCLUSIONS

Trend surface maps

Inspection of Fig.2 reveals four separate regions of erosion where the age of the surface sediment exceeds 1 m.y. on average. A fifth area of scour not revealed in the trend surface analysis covers much of the Campbell Plateau, south of New Zealand (Fig.1a). Extensive erosion on this feature has been inferred from deep-sea drilling and the nature of local seismic profiles (Kennett et al., 1975) A paucity of piston cores from this area (Fig.1b) precludes detecting it, however, as an area of erosion from the trend surface analyses. The most intense area of scour appears to be in the Ross Sea region, the effects of which have been described by Fillon (1975). Scour regions centered on the Kerguelen Plateau and the South Tasman Basin and western Emerald Basin are more extensive, but are on average less eroded than the Ross Sea feature. Another zone of scour is centered in the northwestern sector of the South Australian Basin. The previously published trend surface maps of the western and eastern parts of the area (Watkins and Kennett, 1972; Kennett and Watkins, 1976) were not of sufficient extent to delineate a broad region of minimal erosion adjacent to the Antarctic continent centered west of longitude 120° E and extending to the northern flanks of the Southeast Indian Ridge. This region separates three of the five scour regions.

The configuration of the trends of the thickness of sediment deposited during the past 0.7 m.y. (Fig.3a), reveals not surprisingly, the same general features as those based on the age data (Fig.2a) but some refinements may have emerged. (1) Effects of the scour activity responsible for the hiatus centered on the South Tasman Basin appears to begin no further west than longitude 120° E, and to extend no further east than longitude 190° E. (2) The Kerguelen Plateau scour region has slight west to east elongation in Fig.3. (3) The saddle between these two scour regions features low gradients which steepen northwestwards and northeastwards towards regions of higher sedimentation rates. Thus the saddle between the Kerguelen Plateau and

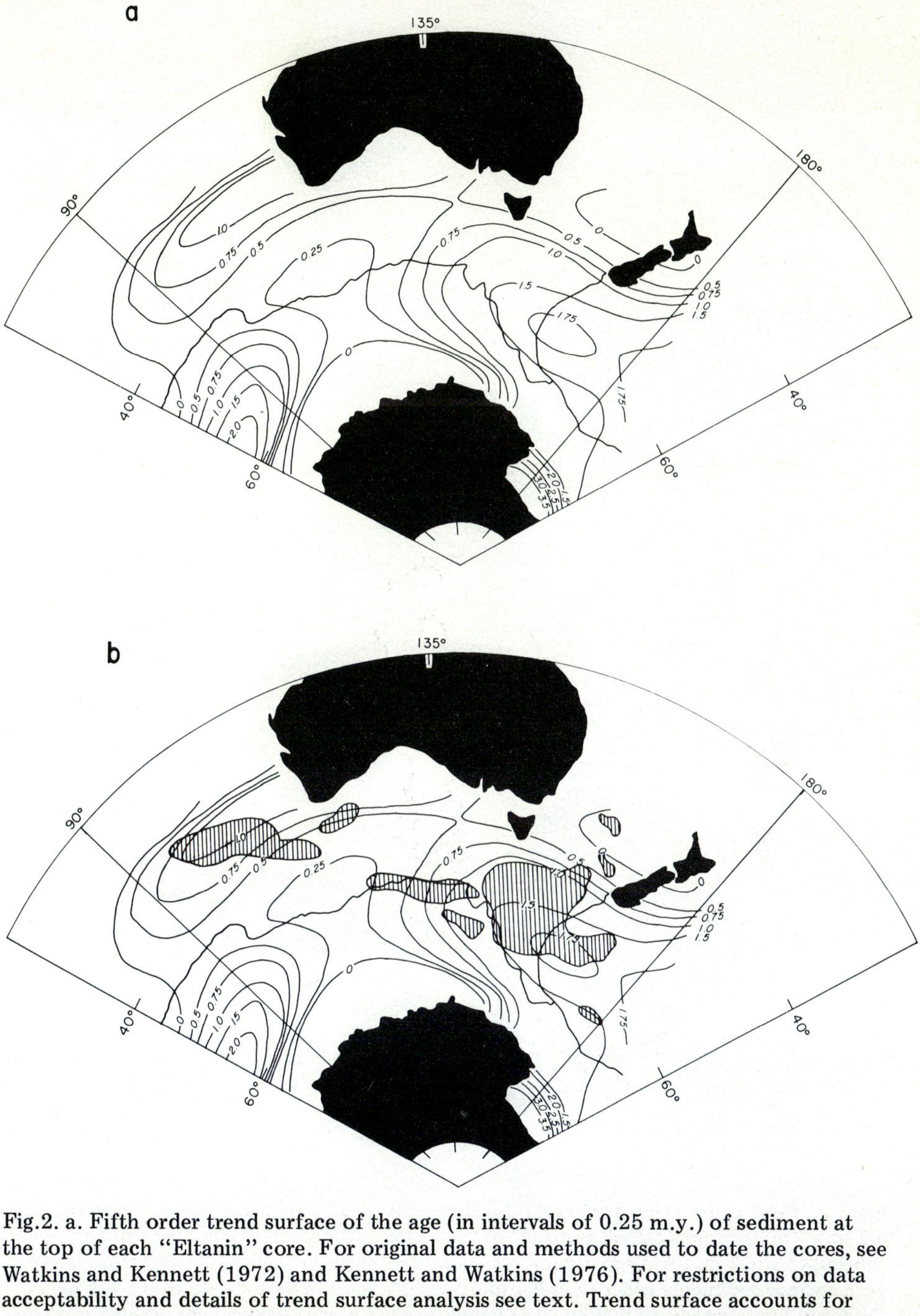

Fig.2. a. Fifth order trend surface of the age (in intervals of 0.25 m.y.) of sediment at the top of each "Eltanin" core. For original data and methods used to date the cores, see Watkins and Kennett (1972) and Kennett and Watkins (1976). For restrictions on data acceptability and details of trend surface analysis see text. Trend surface accounts for 68% of the total data variability. b. As in a, with manganese nodule distribution added, according to Goodell (1973) and Kennett and Watkins (1976) based on bottom photograph observations. Shown only are fields where manganese nodules are common to abundant or where they form a pavement.

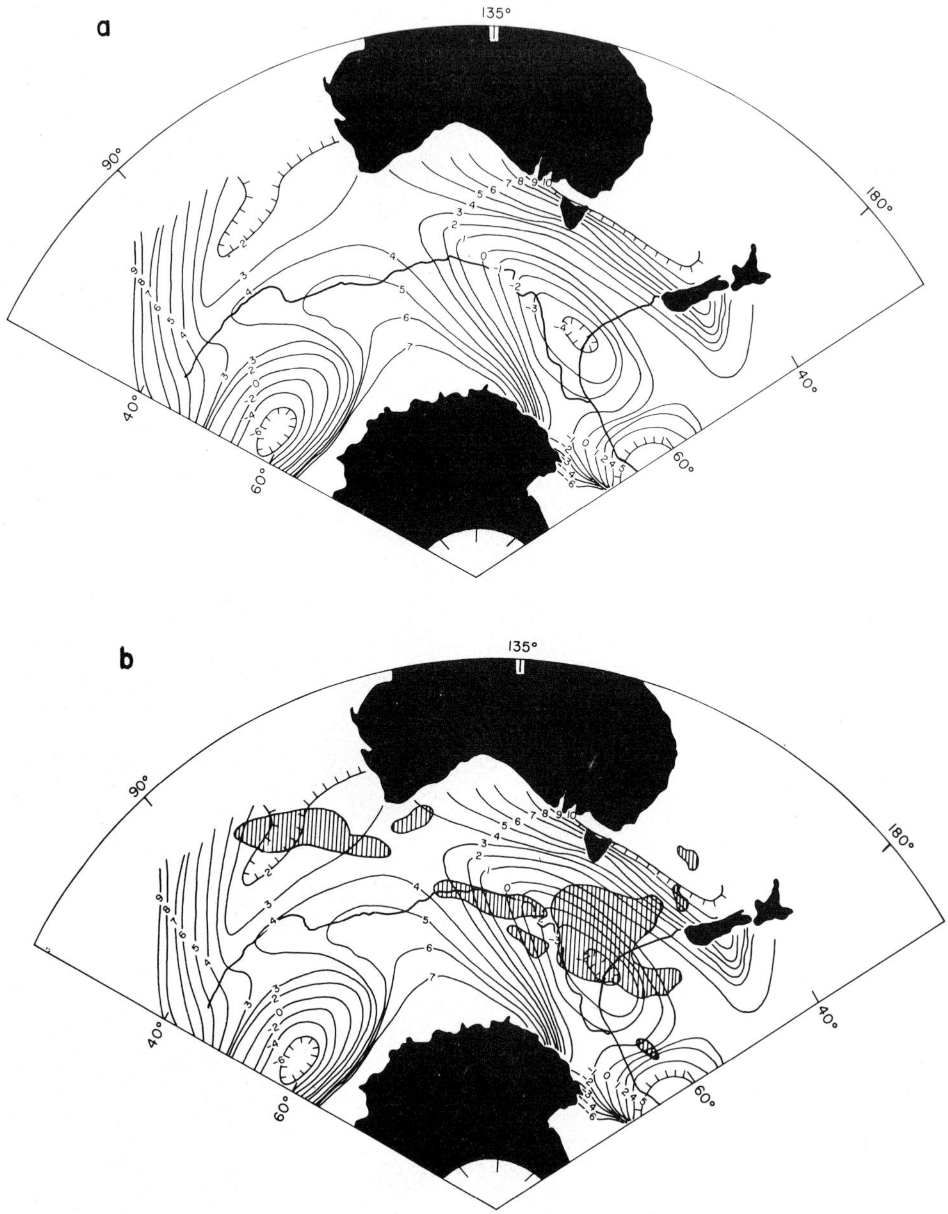

Fig.3. a. Fifth order trend surface of the thickness of sediment deposited during the past 0.7 m.y. (the Brunhes Epoch isopach) in the "Eltanin" piston core collection. See caption to Fig.2. a, for other comments. Note that negative values (derived from cores with surface sediments older than 0.7 m.y., using a sedimentation rate of 0.5 cm per 1000 yrs.) correspond to regions of erosion. b. As in a, with manganese nodule distribution added according to Goodell (1973) and Kennett and Watkins (1976). See caption to Fig.2.b for other comments.

South Tasman Basin scour zones is not so much a higher sedimentation rate feature, as one of moderate but relatively constant sedimentation rate.

Manganese nodule distribution

Inspection of Figs.2b and 3b shows that the area of major manganese nodule development is closely associated with the South Tasman Basin scour area. This has been termed the Tasman Manganese Pavement. A narrowing of manganese nodule fields towards the west follows the trend of the isopachs. The only other major area of manganese nodule development is centered in the northwestern sector of the South Australian Basin, and termed the Southeast Indian Ocean Manganese Pavement (Kennett and Watkins, 1976). This is clearly restricted to that part of the South Australian Basin marked by erosion and slow rates of deposition.

Conditions favoring manganese nodule development

It is well known that manganese nodules often occur in areas marked by strong, sediment-eroding bottom currents or where the rate of sediment deposition is sufficiently low to ensure that accreting manganese does not become covered and removed from the sediment-water interface (Goodell, et al., 1971; Watkins and Kennett, 1971, 1972; Morgenstein, 1972, Glasby, 1972; Horn et al., 1972; Lonsdale, et al., 1972; Kennett and Watkins, 1975, 1976). This is particularly pronounced in those high latitudes of the Southern Hemisphere which are swept by strong bottom currents associated with the Antarctic circum-polar current. In the Australasian sector of the Southern Ocean less than 15% of the manganese nodule fields occur in regions where the average Brunhes isopach exceeds 2 m (Fig.3b), or where the age of the surface sediment is less than 0.5 m.y. All extensive fields are clearly associated with erosion of deep-sea sediments and even localized occurrences of manganese nodules are generally associated with ocean floor features clearly indicating bottom water activity (Kennett and Watkins, 1976).

On the other hand, all areas where extensive erosion has occurred are not associated with widespread manganese nodule development. Sediment surfaces exceeding 1 m.y. in age are common on the Kerguelen and Campbell Plateaus, and in the Ross Sea (Fig.2b), and these are not marked by extensive manganese nodule development. It is therefore clear that while erosion enhances manganese nodule development, such dynamic activity by itself is not sufficient to induce widespread manganese growth. Other factors considered of importance in the development of manganese nodules involve the chemical environment and the proximity of element sources (Mero, 1960).

The Ross Sea and the Kerguelen and Campbell Plateaus are substantially different from the South Tasman Sea and the northwestern sector of the South Australian Basin in two respects: water depth and tectonic setting. Water depth or ambient sediment type alone clearly play no major role in controlling manganese nodule development (Morgenstein, 1972). It is suggested that the tectonic setting is likely to be a significant variable. The Kerguelen Plateau is a classical aseismic ridge, the Campbell Plateau is the southern extension of the New Zealand continental plateau, and the Ross Sea has no notable tectonic association. In contrast, the South Tasman manganese pavement appears closely associated with several major mid-oceanic tectonic elements (Fig.1), and the Southeast Indian Ocean Manganese Pavement is adjacent to the Diamentina Fracture Zone, the western end of which intersects Ninety East Ridge (Fig.1), which is a major transform fault according to McKenzie and Sclater (1971). The proximity of volcanically active sea floor which acts as an element source, is therefore a probable explanation of extensive manganese nodule growth in areas of erosion. On the Campbell Plateau, the only known occurrences of manganese nodules are localized in the vicinity of the Antipodes Islands, a center of late Tertiary of Quaternary volcanism. Initially, this lead Summerhayes (1967, 1969) to suggest that availability of elements from local volcanic sources was essential for manganese growth, and that apparently similar areas of erosion or slow deposition elsewhere on the Campbell Plateau lacked manganese growth because of lack of local submarine volcanism. This interpretation was later superseded by one that invokes differences in the degree of oxygenation of bottom waters (Glasby and Summerhayes, 1975). It was suggested for instance, that, compared with the eastern Campbell Plateau, conditions on the western section were apparently never sufficiently oxidising to allow manganese oxide precipitation. That such differences do exist in the oxidizing activities of the respective bottom waters has, however, not yet been demonstrated. In the South Pacific sector of the Southern Ocean, Goodell (1973) suggested a close association between manganese nodule development, current activity, and submarine volcanism of the mid-oceanic ridge system, and the Eltanin Fracture Zone.

It is therefore suggested that deposition of manganese in the South Tasman and Southeast Indian Ocean was related to the onset of extensive erosion by Antarctic Bottom Waters (Glasby, 1973) and that element sources necessary for this growth have been supplied by relatively local submarine volcanism. A lack of extensive manganese nodule growth in other areas of erosion may be due to a general lack of nearby submarine volcanism.

ACKNOWLEDGEMENTS

This research has been supported by grants by the Office of Polar Programs, National Science Foundation. The most recent grants are OPP75-19222A01 to NDW and OPP75-15511 to JPK.

REFERENCES

Fillon, R.H., 1975. Late Cenozoic paleo-oceanography of the Ross Sea, Antarctica. Geol. Soc. Am. Bull., 86: 839—845.

Glasby, G.P., 1972. The geochemistry of manganese nodules from the northwest Indian Ocean. In: D.R. Horn (Editor). Papers from a Conference on Ferromanganese Deposits on the Ocean Floor, January 20—22, 1972. Arden House/Lamont-Doherty Geological Observatory, pp.93—104.

Glasby, G.P., 1973. Manganese deposits of variable composition from North of the Indian—Antarctic Ridge. Nature Phys. Sci., 242: 106—108.

Glasby, G.P. and Summerhayes, G.P., 1975. Sequential deposition of authigenic marine minerals around New Zealand: paleoenvironmental significance. N.Z. J. Geol. Geophys., 18(3): 477—490.

Goodell, H.G., 1973. Sediments. Am. Geogr. Soc., Antarct. Map Ser., Folio 17.

Goodell, H.G., Meylan, M.A. and Grant, B., 1971. Ferromanganese deposits of the South Pacific Ocean, Drake Passage, and Scotia Sea. In: J.L. Reid (Editor). Antarctic Oceanology, 1. Antarct. Res. Ser., 15: 27—92.

Horn, D.R., Ewing, M., Horn, B.M. and Delach, M.N., 1972. World-wide distribution of manganese nodules. Ocean Ind., 7(1): 26—29.

Kennett, J.P. and Watkins, N.D., 1975. Deep-sea erosion and manganese nodule development in the Southeast Indian Ocean. Science, 188: 1011—1013.

Kennett, J.P. and Watkins, N.D., 1976. Regional deep-sea dynamic processes recorded by late Cenozoic sediments of the southeastern Indian Ocean. Geol. Soc. Am. Bull., 87: 321—334.

Kennett, J.P., Houtz, R.E. et. al., 1975. Initial Reports of the Deep Sea Drilling Project, XXIX. U.S. Government Printing Office, Washington, D.C.

Lonsdale, P., Normark, W.R. and Newman, W.A., 1972. Sedimentation and erosion on Horizon Guyot. Geol. Soc. Am. Bull., 83: 289—316.

McKenzie, D. and Sclater, J.G., 1971. The evolution of the Indian Ocean since the Late Cretaceous. Geophys. J.R. Astron. Soc., 24: 437—528.

Mero, J.L., 1960. Minerals on the ocean floor. Sci. Am., 203: 64—72.

Morgenstein, M., 1972. Manganese accretion at the sediment—water interface at 400 to 2400 meters depth, Hawaiian Archipelago. In: D.R. Horn (Editor), Papers from a conference on Ferromanganese Deposits on the Ocean Floor, January 20—22, 1972. Arden House/Lamont-Doherty Geological Observatory, pp. 131—138.

Summerhayes, C.P., 1967. Manganese nodules from the South-western Pacific. N.Z. J. Geol. Geophys., 10(6): 1372—1381.

Summerhayes, C.P., 1969. Marine Geology of the New Zealand subantarctic sea floor. N.Z. Dep. Sci. Ind. Res. Bull., 190.

Watkins, N.D. and Kennett, J.P., 1971. Antarctic bottom water: major change in velocity during the Late Cenozoic between Australia and Antarctica. Science, 173: 813—818.

Watkins, N.D. and Kennett, J.P., 1972. Regional sedimentary disconformities and Upper Cenozoic changes in bottom water velocities between Australasia and Antarctica. Antarct. Res. Ser., 19: 273—294.

Watkins, N.D. and Kennett, J.P., 1973. Response of deep-sea sediments to changes in physical oceanography resulting from separation of Australia and Antarctica. In: D.H. Tarling and S.K. Runcorn (Editors), Implications of Continental Drift To The Earth Sciences. Academic Press, New York, N.Y., pp.787—798.

Watkins, N.D. and Self, R., 1971. An examination of the "Eltanin" dredged rocks from the Scotia Sea. In: J.L. Reid (Editor), Antarctic Oceanology, 1. Antarct. Res. Ser., 15: 331—343.

Watkins, N.D. and Self, R., 1972. A description of the "Eltanin" dredged rocks from high latitudes of the South Pacific. In: R.J. Adie (Editor), Proc. Second Conference on Antarctic Geology and Geophysics, pp.61—70.

Marine Geology, 23 (1977) 113—132

CONTRASTS BETWEEN THE BRUNHES AND MATUYAMA SEDIMENTARY RECORDS OF BOTTOM WATER ACTIVITY IN THE SOUTH PACIFIC

T.C. HUANG and N.D. WATKINS

Graduate School of Oceanography, University of Rhode Island, Kingston, R.I. 02881 (U.S.A.)

(Received July 1, 1976)

ABSTRACT

Huang, T.C. and Watkins, N.D., 1977. Contrasts between the Brunhes and Matuyama sedimentary records of bottom water activity in the South Pacific. Mar. Geol., 23: 113—132.

Variation of textural and compositional parameters in six dated "Eltanin" sediment cores reveals that the velocity of bottom waters in the South-Central and Southwest Pacific has fluctuated much less during the past 0.7 m.y. than in the preceeding 1.7 m.y., when the creation of up to three widespread hiatuses or the persistence of periods of substantially diminished sedimentation rates were the response to pulses of increased bottom current activity. Polynomial regression analysis between diverse constituents and skewness of the sediments shows that a skewness of —0.2 separates two depositional environments: those dominated by high bottom currents (skewness values more positive than —0.2) and those influenced mainly by pelagic settling (skewness values more negative than —0.2). Measurement of skewness in sediments appears to offer a promising means to finely delineate temporal changes in the bottom dynamic processes of the region.

INTRODUCTION

Recognition about two decades ago of ripple and scour marks on sediment surfaces of many parts of the deep-sea floor encouraged attempts to relate bottom current velocity variation to the observable morphological differences (Menard, 1952; Heezen et al., 1959; Hurley and Fink, 1963; Heezen and Hollister, 1964; Heezen et al., 1966a, b, 1968; Fox and Heezen, 1968; Laughton, 1968; Heezen and Johnson, 1969; Harms, 1969; Neumann and Ball, 1970; Allen, 1970; Johnson, 1972; Heezen and Tharp, 1972). Most of the inferences which can be drawn about net bottom current velocities from deep-sea photographs are at best semiquantitative.

Since Krumbein and Aberdeen (1937) related the grain size distribution of sediments in shallow waters to the changes in bottom water turbulence, well over 100 papers have been published on this general topic. The principle involved is that with increase in velocity, currents will affect sediments so that the mean grain size shifts from fine to coarse; sorting changes from poor to good; and skewness changes from negative to more positive. Harms (1969) has argued that the hydraulic conditions of deep water traction transport must be similar to those in shallow water environments. If this assumption is correct, it becomes obviously possible in principle to indirectly monitor bottom current variations in deep-sea sediments using sediment textures. No such attempt has so far been made.

Recent recognition of widespread erosional disconformities recorded in both piston and Deep-Sea Drilling Project cores (Heezen et al., 1966a; Watkins and Kennett, 1971, 1972; Kennett et al., 1972, 1975; Davies et al., 1975) has given new emphasis to the possible significance of long-term changes in bottom water velocities. Identification of hiatuses has been based primarily on paleontologic and paleomagnetic studies of the cores, which limit the recognition of hiatuses to those exceeding 0.2 m.y. or more (Watkins and Kennett, 1971, 1972).

Methods which can be used to infer the presence of a hiatus (but not measure the time involved) include use of the presence of manganese nodules and coarse materials (Allen, 1970; Watkins and Kennett, 1971, 1972, 1973) and bedding feature analysis using X-ray radiography (Huang and Goodell, 1970; Huang and Stanley, 1972). A lack of manganese nodules or coarse materials does not, however, exclude the possibility that fast bottom currents have occurred (see, for example, Watkins and Kennett, 1977). In general, hiatuses will provide only extreme limits on the age of the bottom water pulses involved in their creation, since by definition, the erosion will remove sediment column containing information about precursors to the scouring activity, and (depending on compaction) perhaps much earlier activity. Full temporal variations cannot therefore be deduced from cores with hiatuses. To detect disconformities spanning periods shorter than 0.2 m.y. special techniques are required. Similarly, since substantially diminished sedimentation rates must exist on the flanks of regions where scour is active (since fast eroding currents must grade laterally into slower currents which merely remove only the finer part of any deposit), definition of the full *spatial* range of any dynamic effect of bottom currents on deep-sea sediments also requires development of means to recognize dynamically imposed lowering of sedimentation rates.

In this paper we shall describe attempts to employ textural and compositional analyses to detect short periods of erosion and diminished sedimentation rates in deep-sea sediments from the Southern Ocean. We shall also search for a single textural parameter which can be used to differentiate quantitatively between different dynamic regimes on the sea floor.

SEDIMENT DISTRIBUTION AND HYDROGRAPHY

The surface sediment distribution in the Southern Ocean has been described by Goodell (1973), who summarized the previous studies, and subdivided the sediments into five major types: calcareous ooze; calcareous—siliceous ooze; siliceous ooze; clayey silt and silty clay; and shelf and coastal debris. Except for the last type, these sediments are commonly found in the "Eltanin" cores retrieved from the Pacific sector of the Southern Ocean (Goodell and Watkins, 1968; Huang et al., 1975a). The present distribution of surface sediments in the Antarctic Ocean (Goodell, 1973, plate 2) is controlled by two main factors, which are the circulation of the Antarctic Ocean and position of the ice-pack fringing the continent (Hays, 1967).

The prevailing westerly wind north of latitude 65°S drives the Antarctic surface water to the east with a northerly component (Fig.1). This water sinks at the Antarctic Convergence, and continues northwards at depth as Antarctic Intermediate Water. Near the Antarctic Continent water sinking also occurs, leading to the formation of Antarctic Bottom Water which flows to the east and north (Gordon, 1967, 1971). In the Pacific sector of the Antarctic Ocean, the Ross Sea is probably the primary source region for the formation of this bottom water (Jacobs et al., 1970).

The high production of siliceous planktonics in the Antarctic Surface Water results in a belt of siliceous ooze around the Antarctic Continent with its southern and northern limits roughly corresponding to the northern and southern limits of the Antarctic Divergence and Convergence, respectively

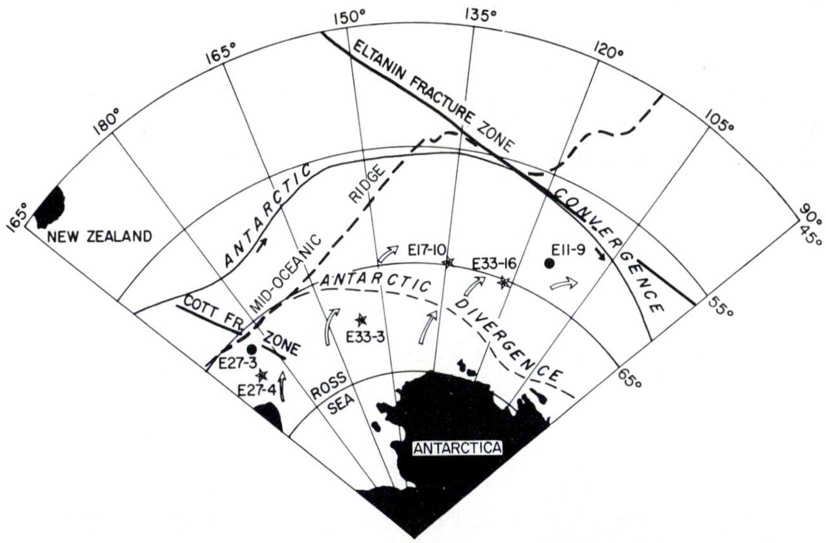

Fig.1. Map showing location of six sediment cores examined. Core sites marked as stars indicate presence of Lower Matuyama hiatuses. Present circum-Antarctic surface flow directions shown as solid arrows; possible bottom current flow direction shown as open arrows. The present position of the Antarctic Convergence, Antarctic Divergence, the "Eltanin" Fracture Zone, and other major tectonic elements from Heezen and Tharp (1972) and Goodell (1973).

(Hays, 1967; Goodell, 1973, plate 2). The sediments south of the siliceous belts are mostly glacial marine (Philippi, 1910) and the sediments north of the Convergence are predominantly calcareous ooze.

METHODS

Laboratory

The six cores examined (Fig.1, Table I) have been dated using paleomagnetic and micropaleontologic methods, and the results have been published in detail elsewhere (Hays, 1965; Watkins and Goodell, 1967; Goodell and Watkins, 1968; Kennett and Watkins, 1970; Watkins and Kennett, 1971, 1972). Subsequently, the volcanic ash distribution in the cores was incorporated to refine the detailed stratigraphy (Huang et al., 1975a). The methods involved in volcanic ash separation and counting, and particle size and compositional analyses have been described by Huang et al. (1975b). A sample interval of 10 cm has been employed in all six cores examined. Particle size analysis of all samples was made using an automated Cahn Sedimentation Balance, and standard sieve methods. X-ray radiographs were made of all cores.

Textural analyses and statistical treatment

Textural parameters, including mean grain size, sorting, skewness, and kurtosis were determined using the moment functions of Blatt et al. (1972). Percent sand and percent silt were also measured. As summarized in Table II, polynomial regression analysis through the 4th degree and analyses of variance (f = 95%) were used so that any compositional factors controlling variation in skewness could be identified. The full range of compositional

TABLE I

Location, water depth and net sedimentation rates of "Eltanin" cores analysed

Core No.	Latitude (S)	Longitude	Water depth (fm.)	Core length (cm)	Net sedimentation rates (cm/1000 yrs)	
					Brunhes	Matuyama
E27-4	68°02.9′	174°35.5′E	1886	1860	0.97	0.91
E27-3	66°04.0′	176°30.0′E	1934	1398	0.71	0.51
E33-3	68°03.2′	159°48.6′W	1940	903	1.15	—
E17-10	65°01.0′	134°52.0′W	2451	1540	0.58	0.67
E33-16	63°14.2′	120°01.2′W	2687	1507	0.69	0.43
E11-9	62°41.0′	115°04.0′W	2740	1700	1.38	0.65

Brunhes refers to the Brunhes normal polarity Epoch (t = 0.69 m.y. to present); Matuyama refers to the reversed polarity epoch (t = 2.41—0.69 m.y.).

TABLE II

Summary of compositional and textural parameters used to understand source of variation in skew, and the statistical treatment employed

Equation: $Y=A+B_1X+B_2X^2...+B_KX^K$
Independent variable, Y = skewness
Dependent variables, X, include:

I. *Compositional parameters:*
- (1) siliceous planktonics (>43 μm)
- (2) manganese micronodules (>43 μm)
- (3) ice-rafted debris (>88 μm)
- (4) siliceous planktonic fragments* (43—2 μm)
- (5) fine terrigenous minerals (43—2 μm)
- (6) volcanic glass (88—11 μm)

II. *Textural parameters:*
- (7) mean grain size (φ)
- (8) sorting (φ)
- (9) kurtosis
- (10) percent sand
- (11) percent silt

Output: mean of each variable
correlation coefficient (r)
F-ratios
regression equations, through K = 4

* Includes diatoms smaller than 43 μm.

and textural parameters employed are given in Table II. A polynomial regression analysis program (BMD05) was modified to facilitate simultaneous analysis of eleven dependent variables (Table II).

RESULTS AND DISCUSSION

Stratigraphy

The stratigraphy of the six "Eltanin" cores (Fig.1, Table I) is given in Fig.2. Several hiatuses are recognized during the Matuyama Epoch (t = 2.41—0.69 m.y.). Ash layers below the "Eltanin Ash" in E27-3 (Huang et al., 1975a) have provided definitive means to identify these hiatuses across the region. Based on the independent dating methods, the sedimentation rates (calculated for core segments exclusive of those including hiatuses) during the Brunhes Epoch (t = 0.69 m.y.—present) were as much as twice those during the Matuyama, as summarized in Table I. Such a contrast in Brunhes and Matuyama Epochs sedimentation rates would appear to be characteristic of much of the Pacific sector of the Southern Ocean (e.g. Watkins and Goodell, 1967). Hitherto, these temporal differences in sedimentation rates in the Southern Ocean have been attributed to planktonic productivity changes and to ice-rafted debris input variation (Goodell and Watkins, 1968; Warnke, 1970; Goodell, 1973; Gram and Warnke, 1974). X-ray radiograph examination of minor sedimentary structures in the cores shows that multiple sedimentary processes including differential pelagic settling, turbidity and bottom currents, and bioturbation may have played a role in the measured sedimentation rate variations. This conclusion is based on results obtained using criteria described for other regions by Bouma (1964),

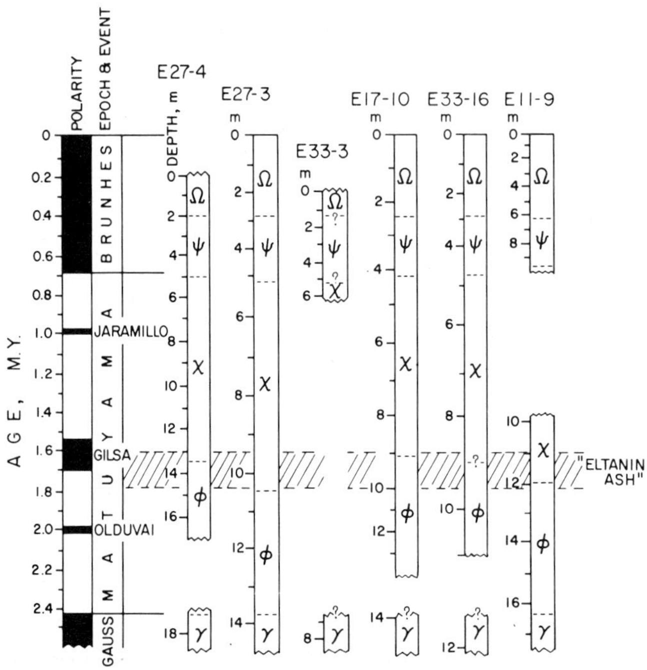

Fig.2. Stratigraphy of six selected cores. Magnetic stratigraphic terms after Cox (1969), using data from Goodell and Watkins (1968) and Kennett and Watkins (1970). Radiolarian zones after Hays (1965); the "Eltanin" Ash horizon from Huang et.al., (1975a).

Stanley and Kelling (1967), Huang and Goodell (1970), and Huang and Stanley (1972). Of all six cores examined only one has a turbidite present, and this is Core E11-9 (Fig.2). Huang et al. (1975a, fig.5) have shown that this was emplaced during the late Matuyama and is probably only a local phenomenon. Hiatuses are also present at the top of cores E27-4 and E33-3, possibly the result of sea floor dynamic processes, but also possibly the result of the coring process itself.

While X-ray radiographic analysis of the cores permits recognition of some extreme dynamic processes, unambiguous recognition of the causes of the substantial temporal variations in sedimentation rates requires statistical analyses of selected textural and compositional parameters, as discussed below.

Textural variations

Skewness, mean grain size and sorting in Core E11-9 (Fig.3) are typical among the six cores examined. It is clear that strong fluctuations exist in the early Matuyama sediments. The major sedimentary processes indicated on the right hand side of Fig.3 were inferred from minor sedimentary structures revealed by X-ray radiographs of the cores. These structures include pelagic

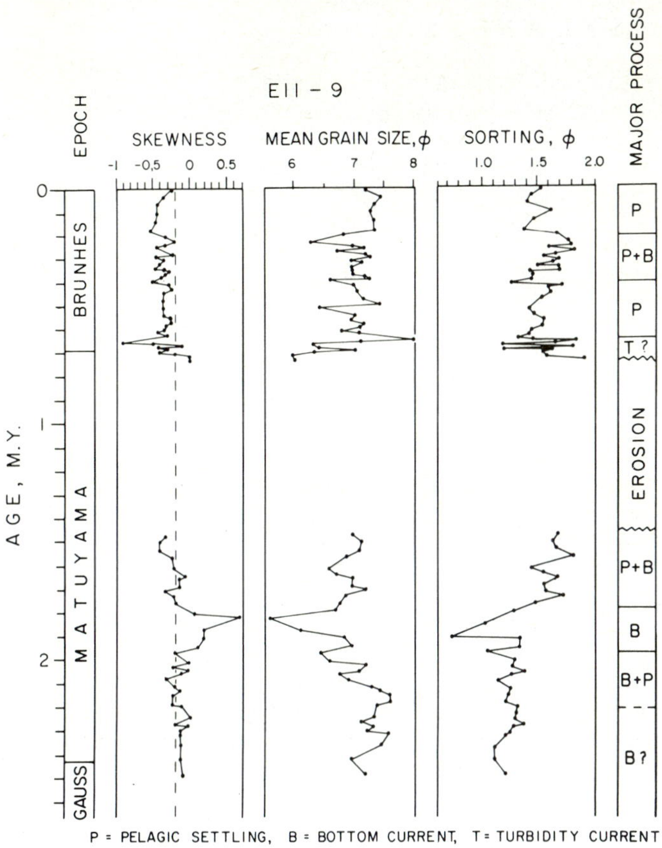

Fig.3. Skewness, mean grain size and sorting variations in core Ell-9. Paleomagnetic age (left) is from Fig.2. Major process is based on analyses of the minor sedimentary structures as revealed by X-ray radiographs.

settling with no features or scattered spots due to presence of ice-rafted debris; fine laminae resembling ripple mark structures caused by bottom current variation; and fine to thick laminae resembling partial Bouma structural sequences, formed during turbidity flow (Huang et al., 1975a).

Fig.4 shows skewness variation in three cores. Brunhes sediment skew values are less variable than those for the early Matuyama, and are almost invariably more negative than —0.2. A sharp positive skewness corresponding to coarser mean grain size and better sorting occurs within the 2.0—1.8 m.y. segments of the three cores (Figs.3 and 4). Lithological details of the cores (Goodell, 1965, 1970; Frakes, 1971) show that silts and sands, often with mottled structures are associated with positive skewness. In contrast, clay or siliceous ooze characterize core segments with negative skewness. The core locations (Fig.1) show that the processes involved span at least 19° of longitude. Extreme positive skewness is to be expected when the finer

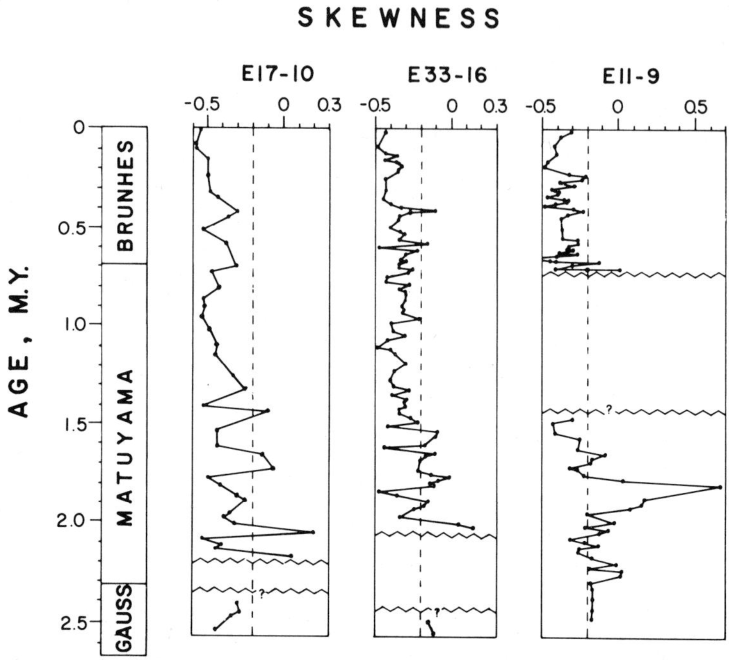

Fig.4. Skewness variation in three selected cores (locations in Fig.1). Note the larger skewness fluctuation associated with the hiatuses (given as data gaps) and the general lack of positive skewness or lack of inferred bottom activity in the Brunhes Epoch.

components of the original sediment are removed by bottom currents, and it is therefore not surprising that the hiatuses in the cores occur in close association with periods of positive skewness. Fig.4 also shows that at least one and possibly three major discrete pulses of increased bottom water velocity occurred during the 2.4—1.7 m.y. period, while bottom waters during the last 0.69 m.y. would appear to have been relatively quiescent in the region. Examination of X-ray radiographs of the sediments characterized by high positive skewness just above the observed hiatuses indicates the presence of lag or residual deposits. In most of these residual deposits laminae resembling ripple mark structures show that lateral traction transport might have occurred during deposition.

This simple interpretation, if based on measurement of skewness alone, would be subject to doubt since other factors can readily effect textural parameters. These include compositional variation, such as those which can be resulted from ice-rafted debris or siliceous planktonic input variation Therefore, differential pelagic contributions might alone, in principle, create a similar textural variation.

Compositional variation

The size distribution of all components in the samples was determined by microscope and scanning electron microscope examination, using conventional counting methods. The major components in the cores include siliceous or carbonate planktonics and their fragments (Fig.5B), ice-rafted debris (Fig.5F), and manganese micronodules (Fig.5D), each with specific and different genetic connotations. Fig.6 shows a representation of the grain size distributions of different components in a typical sample in all the cores. Although the weight percent varies from sample to sample, the particle size range of each component is restricted. All the textural parameters previously computed were based on the total grain size distribution between 250 and 1 μm, which is the sum of the different individual distributions of the subpopulations or components. Sporadic and minor amounts of fractions above 250 μm were excluded from computation within this size range. Variation of any single component could in theory cause a change of total size distribution, and therefore, skewness. Because of this, it is possible that any skew variation caused by fluctuations in bottom water velocity could be obscured by component variations. It is therefore necessary to consider genetic relationships between the constituents, and means of identifying the degree to which they control textural changes. Fig.7 is an example where fluctuations of manganese micronodules and ice-rafted debris accumulation rates vary in harmony, within each of two cores, 3000 km apart (Fig.1). All the six high-latitude cores exhibit similar complementary variations of these two components, and as in Fig.7, feature maxima during the Matuyama Epoch. An increase in these coarse particles causes a shift in skewness from negative towards more positive. There can be little doubt that bottom water activity controls these variations: the ice-rafted debris would tend to be a lag deposit, and the associated decrease in net sedimentation rate would enhance development of manganese micronodules, provided source regions for the manganese are available, and this is probable, in view of the proximity (Fig.1) of the area to various active ridge and fracture elements (Watkins and Kennett, 1977). Surface planktonics can conceivably undergo temporal size variations, and if the bottom sediments were to mirror such an event, skew changes could occur without bottom current activity. Radiolarian tests, mostly larger than 61 μm in diameter, and diatom tests, dominantly *Coscinodiscus* species, which are mostly larger than 43 μm in diameter (Fig.6), would in sufficient volumes cause positive skew if dilution by other fine material or changes during settling do not occur. This is, however, not the case in the present study: in the cores examined the sediments contain up to twelve times more fine siliceous planktonic fragments than unbroken planktonic tests (example in Fig.5B). Fragmentation by bioturbation, and partial dissolution of siliceous plankton on the sea floor and during settling through the water column has been clearly demonstrated elsewhere (Hurd, 1972).

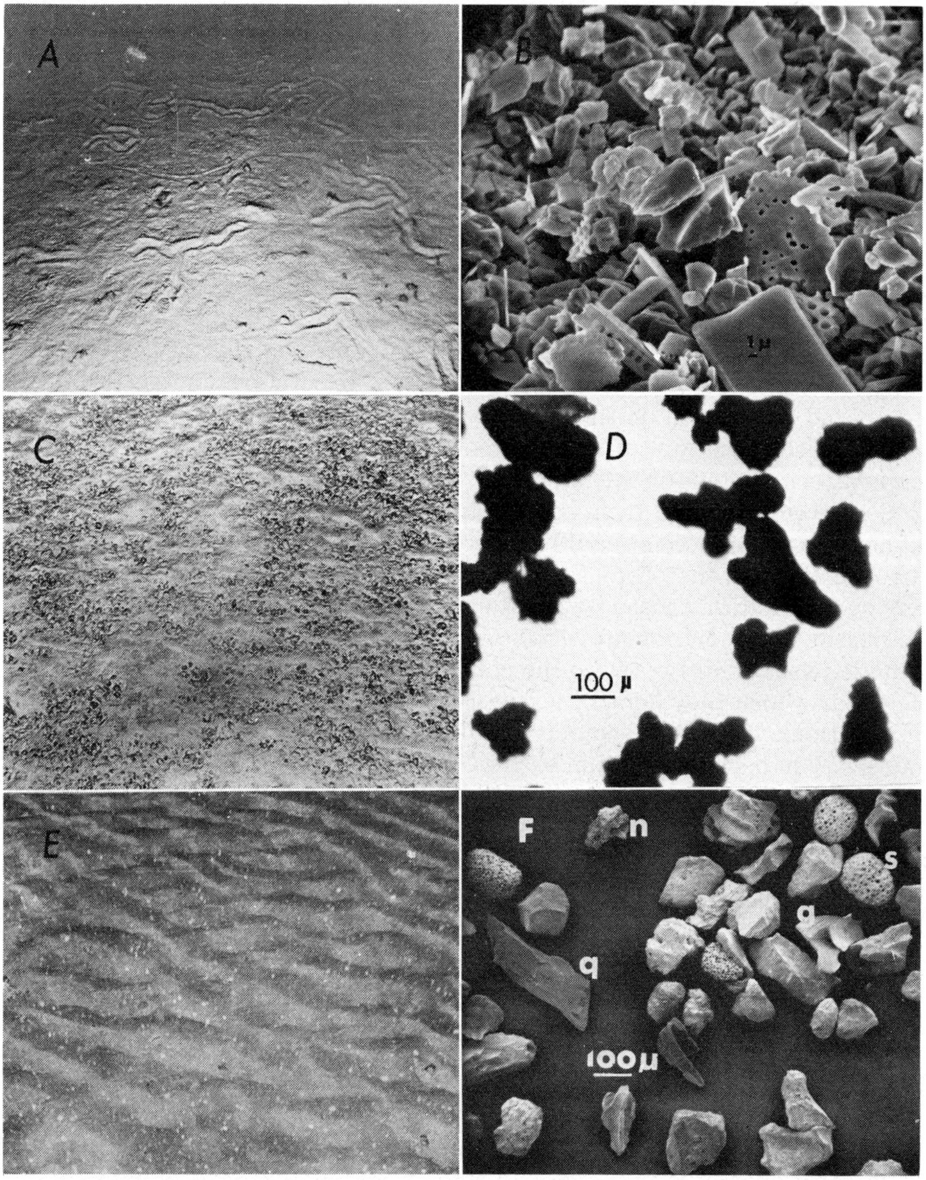

Fig.5. Representative morphologies and associated sediment types in the study area.
A. Photograph showing typical bioturbated mud surface including worm tracks.
B. Scanning electron micrograph showing the siliceous planktonic fragments typical of mud in 5A. (Core E27-4, Table I, depth in core 1360 cm, 43—10 μm size fraction).
C. Photograph showing small manganese nodules and some ripple features (E15-18, 55°58′S 134°26′W, water depth ≈ 3154 m).
D. Photomicrograph of sediment component typical of 5C: manganese micronodules (core E11-9; Table I; depth in core 1357 cm).
E. Photograph showing well-developed ripple marks (E17-1, 55°00′S 135°00′W, water depth = 2941 m).
F. Photomicrograph of sediment typical of 5E: ice-rafted debris including quartz (*q*), glass shards (*g*), radiolarians (*s*), and manganese micronodules (*n*). (E33-3, Table I, depth 455 cm).

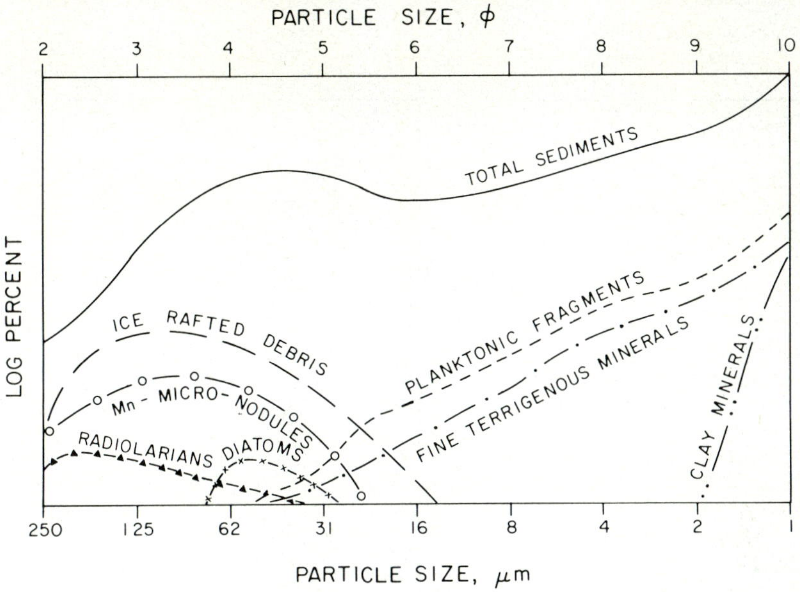

Fig.6. Schematic presentation of size distribution of components in a typical sediment sample. The size range of the individual components is determined using petrographic and a scanning electron microscope.

It is concluded that manganese micronodules and ice-rafted debris can play major roles in skewness variations, and almost certainly strongly reflect variations in bottom water activity. Whether or not changes in other constituents (such as radiolarian and diatom tests) are equally important cannot be argued intuitively, since much depends on the degree of fragmentation occurring penecontemporaneously with deposition, and therefore no generalizations can be made.

Regression analyses

In order to objectively evaluate which component dominates the total size distribution or the skewness variation, polynomial regression analyses of skew versus the apparent accumulation rate of six selected major components and the associated five textural parameters (as summarized in Table II) were made for all the data from three selected cores (E11-9, E33-16 and E17-10). Two series of the analyses were performed: first, the skewness values of the three cores were arbitrarily divided into five groups, with ranges of: more positive than −0.1; −0.1 to −0.2; −0.2 to −0.3; −0.3 to −0.4; and more negative than −0.4. Table III is the result of the regression analysis of these five groups and eleven parameters. All the correlations are expressed in terms of the percent square of correlation coefficients (r^2), which are percent data accounted for by regression. A marked change in percent r^2 values occurs between skewness groups −0.1 to −0.2, and −0.2 to

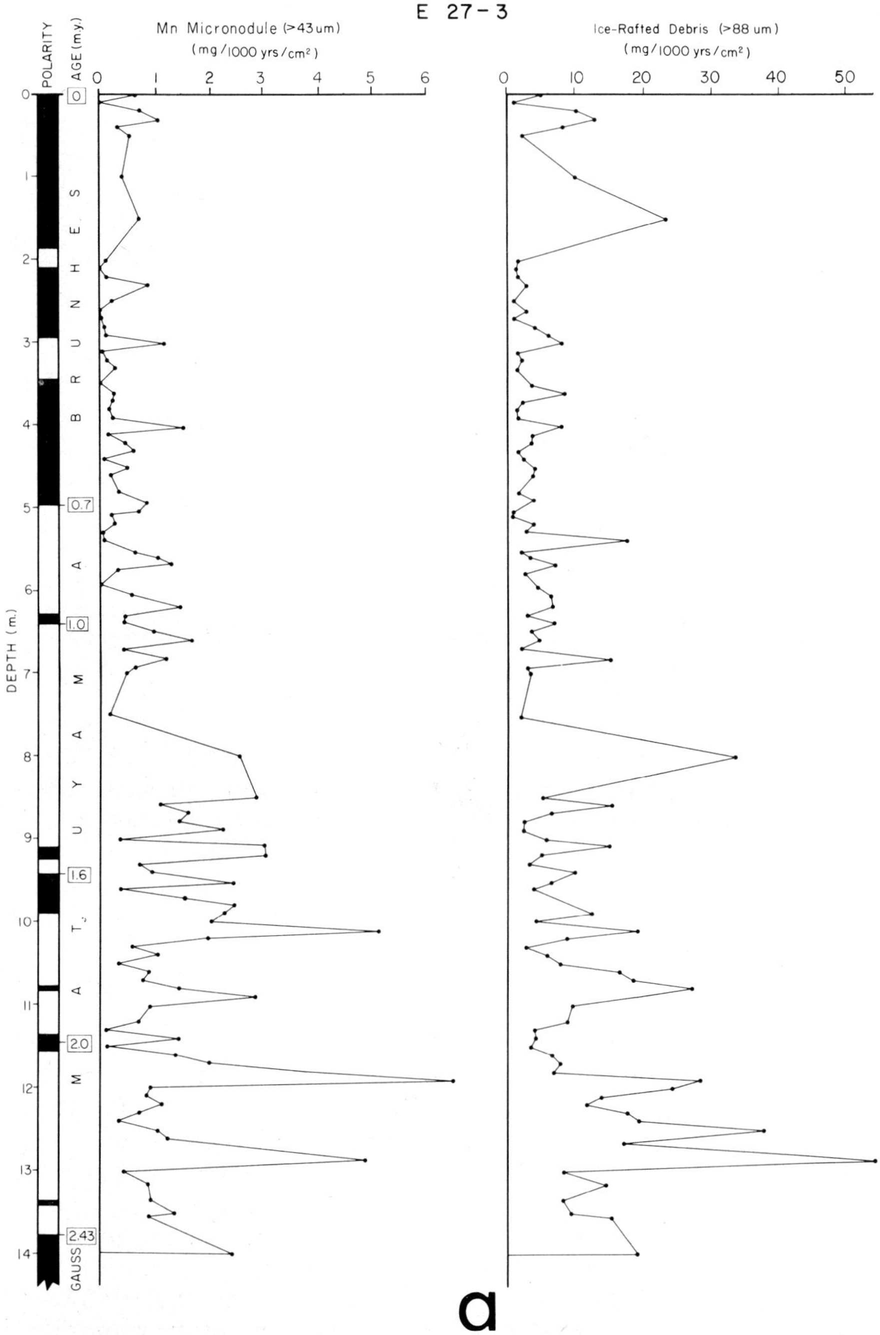

Fig.7. Accumulation rates of manganese micronodules and ice-rafted debris in mg/1000 yrs/cm^2 for (a) core E27-3 and (b) core E11-9, as a function of depth in the core. Ages and magnetic polarity (black = normal, clear = reversed) on left. These data show that the Matuyama epoch is represented by higher apparent accumulation rates of the two

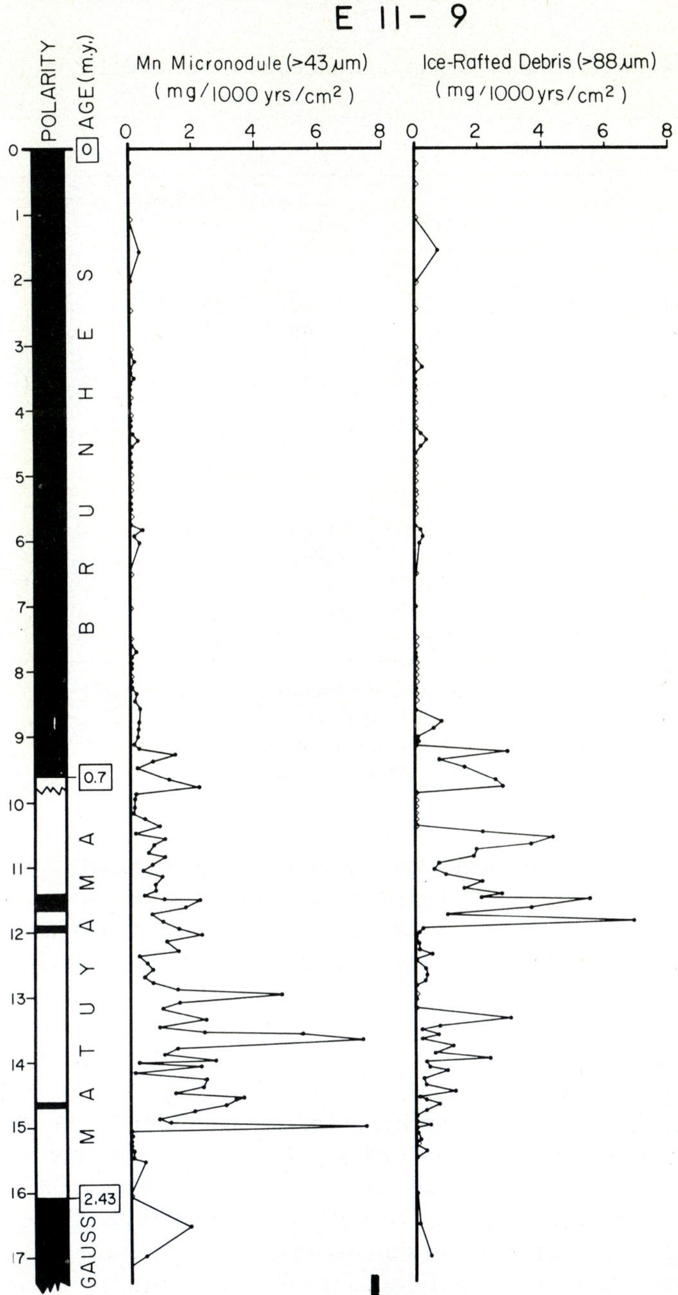

b

independent components in both cores, and are interpreted to show that the ice-rafted debris maxima most probably represent lag deposits, resulting from winnowing, which simultaneously enhances manganese micronodule development.

TABLE III

Correlation† of five skewness groups with composition and texture in cores E11-9, E17-10, and E33-16

Dependent variables	Skewness				
	>—0.1	—0.1 to —0.2	—0.2 to —0.3	—0.3 to —0.4	<—0.4
Compositions					
Siliceous planktonics (whole)	10.35	21.49	15.90	11.11(—)	7.51
Manganese micronodules	48.36*	41.30	5.65	0.31	0.19(—)
Ice-rafted debris	12.09*	69.31*	5.58	7.19	8.52
Fine terrigenous minerals	45.33*	14.64	13.01	8.42	18.39*(—)
Siliceous planktonic fragments	16.74	17.76(—)	29.82(—)	25.92*	38.57*
Volcanic glass	7.17(—)	51.01*	31.11*	2.28(—)	7.53
Textures					
Mean grain size	3.16(—)	1.37(—)	30.53(—)	29.55(—)	0.01(—)
Sorting	9.73(—)	4.96	0.69	0.20	41.20*
Kurtosis	0.14	2.42	5.53(—)	10.73	87.46*(—)
Percent sand	10.89	0.76	18.00	2.31	3.29
Percent silt	5.29	1.76	31.66*	11.07*	30.18*(—)

† Correlation is expressed by percent square of the correlation coefficients (r^2).
* Number which is correlated significantly at 95% confidence level.
(—) Number is negatively correlated (all others are positively correlated).
See Table I for core locations and Figs.2—4 for part of data employed.

—0.3, in that the percent r^2 values of skewness versus manganese micronodule and ice-rafted debris are drastically decreased. Simultaneously, percent r^2 of skewness versus fine siliceous planktonic fragments is increased. On the textural side, percent r^2 of skewness versus mean grain size, sorting, percent sand and silt, also show a marked change between the two groups. Based on this analysis, one can safely conclude that sediments with skewness more negative than —0.2 are significantly different from the sediments with skewness more positive than —0.2.

The second part of the analysis was carried out after dividing the data sets for the three cores into two groups: those with values more negative than —0.2 in the Brunhes Epoch and those with values more positive than —0.2 during the early Matuyama Epoch (t = 2.41—1.70 m.y.). The results (Table IV) show that during the deposition of sediments with skewness more positive than —0.2, manganese micronodules controlled the skewness variation in all the three cores. The next most important contributor to skew variation is ice-rafted debris (Table IV). The reason for its less than dominant role could be that the factors governing ice-rafted debris deposition are not a direct function of bottom current velocity (Watkins et al., 1974; Keany et al., 1976).

TABLE IV

Correlation† of skewness (S) with other parameters in cores E11-9, E17-10, and E33-16

Dependent variables	E11-9		E17-10		E33-16	
	(Lower Mat.) $S>-0.2$	(Brunhes) $S<-0.2$	(Lower Mat.) $S>-0.2$	(Brunhes) $S<0.2$	(Lower Mat.) $S>-0.2$	(Brunhes) $S<-0.2$
Compositions						
Siliceous planktonics (whole)	10.21	50.57*	22.72	14.79(—)	11.40	15.54*
Manganese micronodules	58.20*	0.46	84.13*	0.02(—)	28.03*	2.30(—)
Ice-rafted debris	18.09(—)	6.82	81.59*	1.64(—)	18.77*(—)	3.67
Fine terrigenous minerals	1.83(—)	43.91*	2.89	1.90	3.24(—)	0.17(—)
Siliceous planktonic fragments	0.05(—)	35.28*(—)	23.67	34.53*	1.39(—)	22.99*
Volcanic glass	4.93(—)	3.46	4.02	1.10	4.01	1.11
Textures						
Mean grain size	1.14(—)	43.72*(—)	74.92*(—)	34.96*	1.06(—)	52.39*(—)
Sorting	32.89*	5.37(—)	73.05*	31.64*	14.64(—)	16.67
Kurtosis	7.54	38.09*(—)	62.19*	50.04*	19.45*(—)	68.97*(—)
Percent sand	16.99*(—)	1.57	79.76*	18.61	1.52(—)	23.80*
Percent silt	0.37	39.63*	67.60*	59.30*	21.44*	0.66(—)

† Correlation is expressed by percent square of correlation coefficients (r^2).
* Number which is correlated significantly at 95% confidence level.
(—) Number is negatively correlated (all others are positively correlated).
$S>-0.2$ columns are for data for the Lower Matuyama (t = 2.41 to 1.70 m.y.);
$S<-0.2$ columns are for data for the Brunhes epoch only.
Site location and other details on caption to Table III.

During the deposition of Brunhes sediments with skewness more negative than −0.2, siliceous planktonics and their fragments effected most of the skewness variation. For the textural parameters, sorting controlled most of the variation for those sediments with more positive skewness (>−0.2) and mean grain size, and percent silt controlled most of the skewness variation for sediments with more negative skewness (<−0.2). Manganese micronodules are believed to form authigenetically under high oxidation conditions, favored by extremely slow ambient sedimentation rates, and this condition is associated usually with existence of strong bottom currents (Watkins and Kennett, 1971, 1972, 1977; Kennett and Watkins, 1975). The fact that percent r^2 of sorting and skewness increases for those sediments with skewness more positive than −0.2 supports the existence of bottom current activity during deposition of manganese micronodules.

Based on the above analysis, a skewness value of −0.2 appears to mark the division between those depositional environments influenced by pelagic settling (skewness values more negative than —0.2) and those dominated by bottom current (skewness values more positive than —0.2), and these conclusions have been incorporated into Fig.3. We believe that the positive skewness values in the cores reflect maxima in past bottom current velocity. It is stressed that the relative current velocities inferred from variation in skewness would be net variations using 1400—4600 year signals since all the samples analyzed are 2.0 cm thick, with sedimentation rates ranging from 0.43 to 1.38 cm per thousand years (Table I).

Pulses in bottom current velocity: temporal distribution

The total sedimentological evidence and our statistical analyses indicate clearly that several discrete bottom current pulses of increased bottom current velocity occurred in the Pacific segment of the Southern Ocean during the early Matuyama (Fig.3). It can also be seen that sharp disconformities exist from negative to sharp positive skewness, across disconformities (Fig.4). These changes are to be expected, as a simple model illustrates (Fig.8): as bottom current velocity increases, a critical point (αk) is reached in the skewness. The bottom current then begins to erode (to dashed line depth in Fig.8). This erosion would probably remove positively skewed sediments previously deposited during the time of inhibited

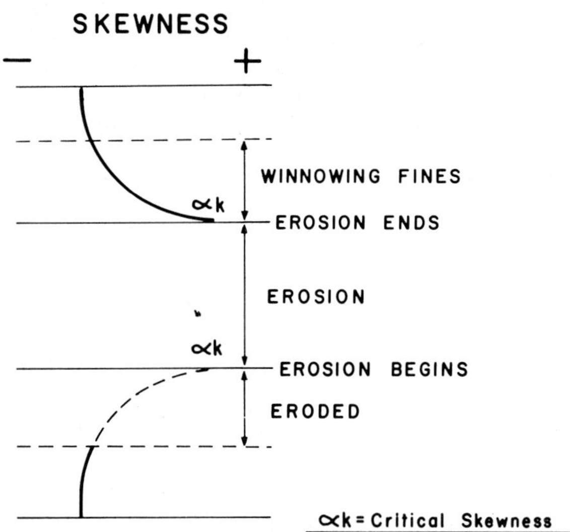

Fig.8. A model relating sediment skewness to erosion by bottom water. An increase in bottom water velocity changes the skew from negative to positive, and decreases the sedimentation rate, until a critical velocity is reached (sediment skew αk) when active erosion begins, removing that sediment deposit during the initial current increase (dashed lines). Erosion ends when the bottom current decreases (skew dropping back to αk), so that the final sampled core will feature a sharp discontinuity in skew (solid curves). Suspected examples of this process exist in the data given as Fig. 4.

deposition (but not erosion). As the current slows (Fig.8, upper), the erosion will cease, and the sediment skew will return towards negative values (Fig.8). The sampled core will then feature a discontinuity from consistent negative to declining positive skewness (solid curve in Fig.8) in association with a hiatus. This sort of analysis could provide a method to estimate the amount of sediments removed during erosion, or to detect otherwise undetectable hiatuses: some may in fact exist at about 1.7 m.y. in E17-10 and E33-16 (Fig.4). The critical skewness (αk) marking the point at which erosion begins, could vary from core to core since the critical bottom current velocity needed for initiating traction transport varies depending on the bedform condition and grain size distribution. This fact may be one reason that the bottom current pulses have caused hiatuses in some cores, but simply substantially reduced deposition rates in others (Table I, Figs.2 and 4). Another reason that erosion and non-erosion contemporaneously exists between cores could be the variation of bottom current velocity between areas. This may be the major reason that the hiatuses and positive skewness peaks are not exactly correlatable (Figs.2 and 4), although there can be little doubt that the same major Lower Matuyama activity has affected all three cores. It may be that during the Matuyama bottom current activity was most vigorous in the area near the Ross Sea as shown by core E33-3 (Fig.1) in which the Matuyama sediments have been almost entirely eroded by bottom currents (Fig.2), as demonstrated by Fillon (1975) for many cores in the Ross Sea area. Other sedimentary processes which can produce hiatuses are turbidity currents. Throughout the six cores examined, only one turbidite sequence has been detected. It occurs in conjunction with a marked hiatus in Core E11-9. The polynomial regression analysis of the constituents and textures of this turbidite shows that it differs from a residual deposit produced by a bottom current activity in that ice-rafted debris rather than manganese micronodules control most of the skewness variation, so that the turbidite is not the product of a prolonged period of scour.

The present distribution of higher Antarctic bottom current velocity coincides roughly to the location of the present Antarctic Convergence (Heezen et al., 1968; Heezen and Tharp, 1972), and is also associated with the present distribution of iron-manganese nodules (Goodell, 1973). Examination of the "Eltanin" bottom photographs for the region reveals that in many respects the present bottom features can be correlated to the sediments: the sea floor environment with little current activity (example in Fig.5A) can be associated with the siliceous planktonic fragments (example in Fig.5B); the morphology characterized by a weak bottom current activity (example in Fig.5C) which conceivably reduces the sedimentation rates may produce the sediments containing abundant manganese micronodules (example in Fig.5D); and the environments characterized by strong bottom current activity with distinct ripple marks (example in Fig.5E) may be analogous to a coarse residual deposit containing ice-rafted debris and manganese micronodules (Fig.5F). Finally, very intense bottom current activity would tend to scour all the sediments and create a pebbly or rocky surface. In fact many bottom photographs beneath the Antarctic Convergence exhibit only large manganese nodules (Heezen and Tharp, 1972).

CONCLUSIONS

We have employed automated systems to derive conventional textural parameters for a series of dated cores from the South Pacific segment of the Southern Ocean. The results show that the Antarctic Bottom Water velocity fluctuated much more during the period 2.4—1.7 m.y. than in the last 0.7 m.y. We can state with certainty that past Antarctic bottom current velocity changes of some substantial but unknown value are unambiguously recorded by skewness variation in the sediments of the Southern Ocean. Short-period disconformities can be detected by dual analysis of skewness variation and compositional parameters. In the region, skewness of —0.2 in sediments separates two depositional environments dominated by the effects of pelagic settling and bottom current activity respectively. At least one major and several minor bottom current pulses have caused hiatuses and residual deposits during the early Matuyama Epoch.

ACKNOWLEDGEMENTS

Supported by the National Science Foundation, Submarine Geology and Geophysics Program grant DES74-22347 and Office of Polar Programs, grant OPP70-02982. "Eltanin" bottom photographs were provided from the Oceanographic Sorting Center of the Smithsonian Institution, Washington, D.C. through Mr. H. Sheng of the Sedimentology Division, Smithsonian Institution.

REFERENCES

Allen, D.W., 1970. Sedimentary Texture — A Key to Interpret Deep-Marine Dynamics. Thesis, Oregon State Univ., Corvallis, Oreg., 152 pp.
Blatt, H.G., Middleton, G. and Murray, R., 1972. Origin of Sedimentary Rocks. Prentice-Hall, Englewood Cliffs, N.J., 634 pp.
Bouma, A.H., 1964. Notes on X-ray interpretation of marine sediments. Mar. Geol., 2: 278—309.
Cox, A., 1969. Geomagnetic reversals. Science, 163: 237—245.
Davies, T.A., Weser, O.E., Luyendyk, B.P. and Kidd, R.B., 1975. Unconformities in the sediments of the Indian Ocean. Nature, 253: 15—19.
Fillon, R.H., 1975. Late Cenozoic paleo-oceanography of the Ross Sea, Antarctica. Geol. Soc. Am. Bull., 86: 839—845.
Fox, P.J. and Heezen, B.C., 1968. Abyssal anti-dunes. Nature, 220: 470—472.
Frakes, L.A., 1971. USNS "Eltanin" core descriptions, Cruises 32 to 45. Florida State Univ. Antarct. Core Facility, Tallahassee, Fla., 105 pp.
Gordon, A.L., 1967. Structure of Antarctic Waters between 20°W and 170°W. Am. Geogr. Soc., Antarct. Map Folio Ser., Folio 6: 10 pp.

Gordon, A.L., 1971. Oceanography of Antarctic waters. In: J.L. Reid (Editor), Antarctic Oceanology, I. Antarct. Res. Ser., 15: 169—203.

Goodell, H.G., 1965. Marine Geology, USNS "Eltanin" Cruises 9—15. Fla. State Univ. Sedimentol. Res. Lab. Contrib., 11: 196 pp.

Goodell, H.G., 1970. USNS "Eltanin" core descriptions, cruises 16 to 27. Florida State Univ. Antarctic Core Facility, Tallahassee, Fla., 206 pp.

Goodell, H.G., 1973. The sediments. In: V.C. Rusnell (Editor), Marine Sediments of the Southern Oceans. Am. Geogr. Soc., Antarct. Map Folio Ser., Folio 17: 1—9.

Goodell, H.G. and Watkins, N.D., 1968. Paleomagnetic stratigraphy of the southern ocean: 20°W to 160°E longitude. Deep-Sea Res., 15: 89—112.

Gram, R. and Warnke, D.A., 1974. New evidence for changing rates of biogenic sedimentation in the southern ocean. EOS Trans. Am. Geophys. Union, 55: 311.

Harms, J.G., 1969. Hydraulic significance of some sand ripples. Geol. Soc. Am. Bull., 80: 363—396.

Hays, J.D., 1965. Radiolaria and late Tertiary and Quaternary history of Antarctic Seas. In G.A. Llano (Editor), Biology of the Antarctic Seas, II. Am. Geophys. Union, 15: 125—184.

Hays, J.D., 1967. Quaternary sediments of the Antarctic Ocean. In: M. Sears (Editor), Progress in Oceanography. Pergamon, New York, N.Y., pp.117—131.

Heezen, B.C. and Hollister, C.D., 1964. Deep-sea current evidence from abyssal sediments. Mar. Geol., 1: 141—174.

Heezen, B.C. and Johnson, G.L., 1969. Mediterranean under-current and microphysiography west of Gibraltar. Bull. Inst. Oceanogr. Monaco, 67: 95 pp.

Heezen, B.C. and Tharp, M., 1972. Morphology of the sea floor. In: B.C. Heezen, M. Tharp and C. Bentley (Editors), Morphology of the Earth in the Antarctic and Subantarctic. Am. Geogr. Soc., Antarct. Map Folio Ser., Folio 16: 10—14.

Heezen, B.C., Tharp, M. and Ewing, M., 1959. The floors of the Ocean, I. The North Atlantic. Geol. Soc. Am. Spec. Pap., 65: 122 pp.

Heezen, B.C., Hollister, C.D. and Ruddiman, N.F., 1966a. Shaping of the continental rise by deep geostrophic contour currents. Science, 152: 502—508.

Heezen, B.C., Schneider, E.D. and Pilkey, O.H., 1966b. Sediment transported by the Antarctic bottom current in Bermuda Rise. Nature, 211: 611—612.

Heezen, B.C., Tharp, M. and Hollister, C.D., 1968. Illustrations of the marine Geology of the Southern Ocean. In: Symposium on Antarctic Oceanography, Santiago, 1968, pp.13—16.

Huang, T.C. and Goodell, H.G., 1970. Sediments and sedimentary processes of eastern Mississippi Core, Gulf of Mexico. Am. Assoc. Pet. Geol. Bull., 54: 2070—2100.

Huang, T.C. and Stanley, D.J., 1972. Western Alboran Sea: Sediment dispersal, ponding and reversal of currents. In D.J. Stanley (Editor), The Mediterranean Sea. Dowden, Hutchinson and Ross, Stroudsburg, Pa., pp.521—559.

Huang, T.C., Watkins, N.D. and Shaw, D.M., 1975a. Atmospherically transported volcanic glass in deep-sea sediments: volcanism in sub-Antarctic latitudes of the South Pacific during the Upper Pliocene and Pleistocene. Geol. Soc. Am. Bull., 86: 1305—1315.

Huang, T.C., Watkins, N.D. and Shaw, D.M., 1975b. Atmospherically transported volcanic glass in deep-sea sediments: development of a separation and counting technique. Deep-Sea Res., 22: 185—196.

Hurd, D.C., 1972. Factors affecting solution rates of biogenic opal in sea-water. Earth Planet. Sci. Lett., 15: 411—417.

Hurley, R.J. and Fink, L.K., 1963. Ripple marks show that countercurrent exists in Florida Straits. Science, 139: 603—605.

Jacobs, S.S., Amos, A.F. and Bruchhausen, P.M., 1970. Ross Sea oceanography and Antarctic Bottom Water formation. Deep-Sea Res., 17: 935—962.

Johnson, D.A., 1972. Ocean-floor erosion in the equatorial Pacific. Geol. Soc. Am. Bull., 83: 3121—3144.

Keany, J., Ledbetter, M., Watkins, N.D. and Huang, T.C., 1976. Diachronous deposition of ice-rafted debris in sub-Antarctic deep-sea sediments. Geol. Soc. Am. Bull., 87: 278—309.

Kennett, J.P. and Watkins, N.D., 1970. Geomagnetic polarity change, volcanic maxima and faunal extinction in the South Pacific. Nature, 227: 930—934.

Kennett, J.P. and Watkins, N.D., 1975. Deep-sea erosion and manganese nodule development in the Southeast Indian Ocean. Science, 188: 1011—1013.

Kennett, J.P., Burns, R.E., Andrews, J.E., Clurkin Jr., M., Davies, T.H., Damitrica, P., Edwards, A.R., Galehouse, J.S., Packham, G.H. and Van der Lingen, G.L., 1972. Australia—Antarctic continental drift, paleocirculation changes and Oligocene deep-sea erosion. Nature Phys. Sci., 239: 51—55.

Kennett, J.P., Houtz, R.E., Andrews, P.B., Edwards, H.R., Gostin, V.A., Hajos, M., Hampton, M., Jenkins, P.G., Margolis, S.V., Ovenshine, A.T. and Perchi-Nielsen, K., 1975. Cenozoic paleo-oceanography in the southwest Pacific ocean, Antarctic glaciation, and the development of the circum-Pacific current. In: Initial Reports of the Deep-Sea Drilling Project, 26. U.S. Government Printing Office, Washington, D.C., pp.1155—1169.

Krumbeim, W.C. and Aberdeen, E.J., 1937. The sediments of Baratoria Bay (La.). J. Sediment. Petrol., 7: 3—17.

Laughton, A.S., 1968. New evidence of erosion on the deep ocean floor. Deep-Sea Res., 15: 21—29.

Menard, H.W., 1952. Deep ripple marks in the sea. J. Sediment. Petrol., 22: 3—9.

Nayudu, Y.R., 1971. Lithology and chemistry of surface sediments in sub-Antarctic regions of the Pacific Ocean. In: J.D. Reid (Editor), Antarctic Oceanology, 1. Antarct. Res. Ser., 15: 247—282.

Neumann, A.C. and Ball, M.M., 1970. Submersible observations in the Strait of Florida: geology and bottom currents. Geol. Soc. Am. Bull., 81: 2861—2874.

Philippi, E., 1910. Die Grundaufgaben der Deutsche Südpolar Expedition, 1901—1903, 2. Geographie und Geologie, 6. Georg Reimer, Berlin, pp. 415—616.

Stanley, D.J. and Kelling, G., 1967. Sedimentation patterns in the Wilmington Canyon. In: Ocean Science and Engineering of the Atlantic Shelf. Trans. Mar. Technol. Soc., pp.127—142.

Warnke, D.A., 1970. Glacial erosion, ice-rafting, and glacial marine sediments — Antarctica and the southern ocean. Am. J. Sci., 269: 276—294.

Watkins, N.D. and Goodell, H.G., 1967. Geomagnetic polarity change and faunal extinction in the Southern Ocean. Science, 156: 1083—1087.

Watkins, N.D. and Kennett, J.P., 1971. Antarctic bottom water: major change in velocity during the late Cenozoic between Australia and Antarctica. Science, 173: 813—818.

Watkins, N.D. and Kennett, J.P., 1972. Regional sedimentary disconformities and Upper Cenozoic changes in bottom water velocities between Australia and Antarctica. Antarct. Res. Ser., 19: 273—294 (edited by D.E. Hayes).

Watkins, N.D. and Kennett, J.P., 1973. Response of deep-sea sediments to changes in physical oceanography resulting from separation of Australasia and Antarctica. In: D.H. Tarling and S.K. Runcorn (Editors), Continental Drift, Sea Floor Spreading and Plate Tectonics — Implication for the Earth Sciences. Academic Press, New York, N.Y., pp.787—797.

Watkins, N.D. and Kennett, J.P., 1977. Erosion of deep-sea sediments in the Southern Ocean between longitudes 70°E and 190°E and contrasts in manganese nodule development. Mar. Geol., 23: 103—111.

Watkins, N.D., Keany, J., Ledbetter, M. and Huang, T.C., 1974. Antarctic glacial history derivation by analysis of ice-rafted deposits in marine sediments — a new model and the initial testing. Science, 186: 533—536.

Marine Geology, 23 (1977) 133—154

EFFECTS OF BIOTURBATION ON SEDIMENT—SEAWATER INTERACTION

D.R. SCHINK and N.L. GUINASSO Jr.

Department of Oceanography, Texas A & M University, College Station, Texas 77843 (U.S.A.)

(Received July 1, 1976)

ABSTRACT

Schink, D.R. and Guinasso Jr., N.L., 1976. Effects of bioturbation on sediment—seawater interaction. Mar. Geol., 23: 133—154.

Large gradients of dissolved silica are common in the uppermost layers of abyssal marine sediments. These gradients can be sustained only by an active, continuous supply of dissolved silica from within the sediments. The flux of dissolved silica is approximately equivalent to the total accumulation rate of all siliceous material (including aluminosilicates) in the sediments; if aluminosilicates were the source of dissolved silica they would have to give up nearly all of their silicon. Since marine sediments commonly show no such depletion in silicon with depth, dissolution of aluminosilicates cannot be the primary source of dissolved silica emanating from the sea floor. The main source of this flux must be particulate biogenous silica that is mixed downward into the more inert sedimentary material and dissolves leaving little residue.

Biological and physical processes stir deep-sea sediments at rates which can be described by vertical mixing coefficients. These coefficients can be evaluated from microtektite redistribution or from the depth profiles of plutonium in sediments. Measured rates range between 0.5 and 400 $cm^2 \, kyr^{-1}$ in the deep sea.

By redistributing reactive particles, this vertical mixing influences the chemical gradients in interstitial waters and the chemical fluxes across the sea-floor boundary. Dissolving particles of silica or calcium carbonate will have more effect on pore water concentrations when they are deeper within the sediment.

Model calculations show the input flux of biogenous particulate silica to be the primary controlling factor on the flux of dissolved silica and on the interstitial profile of dissolved silica. Bioturbation rates and dissolution rates also have an important effect on interstitial profiles and this effect can be estimated although the actual rates are poorly known. Model calculations indicate that the concentrations of dissolved species in pore water are controlled by a dynamic balance between competing processes. Equilibrium reactions between sediments in the upper few meters of the sediment column and their associated pore water are almost irrelevant in establishing the composition of the pore waters.

Silica solubility is only slightly affected by temperature and pressure changes in the deep sea, and abyssal waters are always far from saturation. Accordingly the water composition can have only a minimal effect on dissolution rates of SiO_2. On the other hand, the actual mechanism which prevents dissolution of some siliceous fossils remains unknown. A stagnant boundary layer in water overlying the sea floor will have almost no effect on concentrations or fluxes of silica.

Variations in bioturbation rates or in the flux of biogenous particulate silica to the sediment would affect the balance of concentrations in the pore water profile. Such changes over the last few hundred years would be recorded in the composition of pore waters which are associated with sedimentary particles hundreds of thousands of years old. In this sense the "age" of the pore waters is far less than the "age" of the associated solids in the upper meters of the sea floor.

INTRODUCTION

Sediments accumulating on the sea floor do not simply reflect what is delivered to the bottom of the ocean, rather they represent a residuum — the net result of deposition, decomposition, dissolution, and reprecipitation. Reactions occurring in pore water play an important role in determining what stays deposited and what returns to the ocean cycles. Pore water is much like seawater in most of its chemical characteristics; the composition of overlying abyssal waters influences the composition of pore waters and thus affects reactions with the sediments.

Berger (1968) and others have noted the marked influence of different abyssal waters on carbonate dissolution. Berger (1970) suggested that a similar effect influences silica deposition. But silica differs from carbonate in many important respects. Most significantly silica, unlike $CaCO_3$, is never saturated or even close to saturated in ocean waters; silica solubility decreases as temperature decreases; and silica solubility is not so strongly pressure dependent (Jones and Pytkowicz, 1973; Willey, 1974). Whereas accumulation of carbonate sediments is closely related to solubility equilibria, the real reason for accumulation and preservation of amorphous silica remains something of an enigma.

Processes affecting interstitial silica are easier to study than those affecting carbonates. Carbonate solubility is markedly affected by temperature and pressure, and solubility equilibrium is notoriously slow. Since carbonate solution and precipitation are not readily reversible, and may have very different rates, all analyses of carbonate ion in pore water must be suspect if pore waters have come through temperature and pressure gradients in contact with carbonate solids. On the other hand, such effects on interstitial silica appear to be modest and mostly reversible so that pore waters separated at in situ temperatures seem to give an accurate reflection of their in situ silica content.

Pore waters are linked to the overlying seawaters by the process of aqueous diffusion. It is this process which holds the composition of pore waters close to that of seawater. Any reactions which produce or remove dissolved species will generate chemical gradients, and initiate aqueous diffusion. Diffusion is a relatively rapid process compared to sedimentary accumulation and so some reaction must actively produce or consume dissolved material if a chemical gradient is to be sustained. Over the long and slow sediment-accumulation process even a small gradient can be sustained only by deposition or removal of a large amount of material.

Reactions in the upper meter of marine sediments most strongly control the chemical fluxes across the sea-floor boundary. Over longer length scales, at greater depths, we see other important reactions slowly affecting the final form of buried sediments and their associated brines. These diagenetic reactions play a critical role in establishing the final character of buried material, but they are not so important to sediment—seawater interactions.

Fig.1 shows typical dissolved interstitial silica profiles from different oceans measured by different investigators. Virtually every marine sediment ever examined has shown concentrations of dissolved silica in the interstitial waters significantly higher than concentrations in the overlying seawater (Schink et al., 1974). In every case we find a steep gradient near the interface. Such gradients imply substantial fluxes of silica from sediments to seawater.

Some confusion concerning these fluxes deserves brief comment. It has been argued that concentration gradients in reactive porous media do not necessarily produce diffusive fluxes. Such arguments are wrong, unless the pores do not interconnect; such is surely not the case for abyssal sediments of 60—90% porosity. The presence of non-reactive solid particles slows the diffusive flux by forcing the migrating species into more indirect routes and by reducing the cross-sectional area through which diffusion can proceed. In most marine sediments, with porosities of 0.6—0.9, these two effects combine to slow aqueous diffusion by 30—40%. Reactions between solids and the liquid phase could stop aqueous diffusion only if aqueous gradients were reduced to zero. A reaction, such as adsorption, may pull most of the diffusible material out of solution; the non-adsorbed fraction remaining in

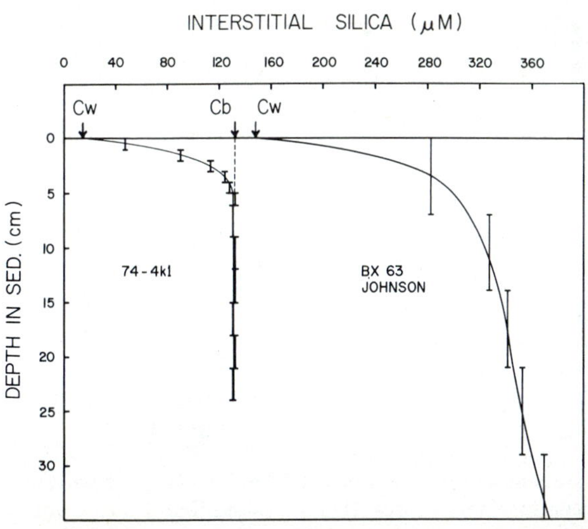

Fig.1. Dissolved silica profiles in the pore waters of two different cores. Core 74—4K1 is a gravity core taken by the authors from 4521 m depth at 20°17′N, 83°36′W. Core BX 63 is a box core taken from 4090 m depth at 14°32′N, 117°15′W as described by Johnson (1975).

solution will still diffuse down its gradient in the aqueous phase just as before. If we somehow imposed a layer of silica adsorbant on the surface of the sediments, this material might react with upward diffusing dissolved silica and drop the concentration gradient to zero inside the adsorbant layer; but below the layer, where the dissolved concentration gradient persisted, there would still be a diffusive flux upward.

Some appreciation of the effect of these gradients might be derived from an evaluation of associated fluxes. If aluminosilicates are accumulating at 0.6 g cm^{-2} kyr^{-1} (equivalent to 1 cm kyr^{-1} at 0.75 porosity, average solids density of 2.6 g cm^{-3}), and if the SiO_2 content of the aluminosilicate material were 50% — which is about normal — then the entire silicon fraction of the accumulating sediment would have to dissolve and diffuse away in order to sustain a dissolved silica gradient of 55 μM cm^{-1}. Gradients of this magnitude are common, although uncertainties in their measurement — produced by sampling, temperature and pressure effects — often confuse the issue. Fig.2 compares gradients in the two profiles from Fig.1 with the interface gradient equivalent to total dissolution of silicon.

In spite of the frequent occurrence of steep gradients, the sea floor is not covered by widespread deposits of pure gibbsite or other silica-free forms. We would argue then that the observed silica gradients and the inferred silica fluxes must be sustained by dissolution of some species which disappears leaving little or no residue. Amorphous, biogenous silica provides such a

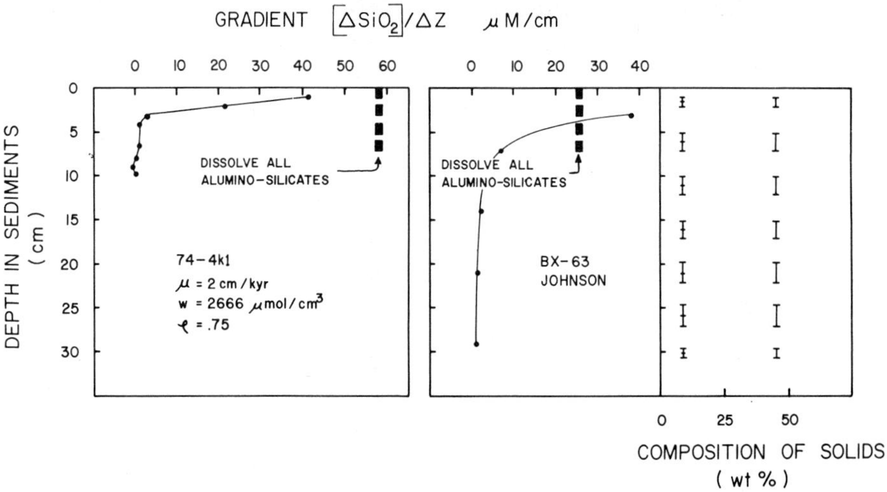

Fig.2. Gradient of interstitial silica concentration as estimated from $\Delta[SiO_2]/\Delta z$ where Δz is based simply on distance between interval centers. Heavy dashed line shows gradient at interface (or below) that would require total dissolution of all silica in accumulating aluminosilicates if they were the sole source of support for the silica gradient. Composition of solids in the Johnson core shows no depletion of silica. (The simple bars represent SiO_2 ; the bars with the center cross represent Al_2O_3.) Cores are same as in Fig.1.

source, but since this material dissolves in times short compared to burial rates it seems, at first, impossible that soluble particulate silica could be found at depth in the sediment. Vertical mixing of solid particles at the sea floor is usually responsible for such burial. Vertical mixing may be accomplished by physical movements of the waters at the sea floor. More important must be the myriad activities of the organisms living on or in the sea floor. They eat, bore through, burrow in, spray out, or otherwise disturb and mix the sediments lying near contact with seawater at the floor of the ocean. The combination of all these actions has been termed bioturbation, and this process has a significant influence on the chemical interactions between sediment and seawater.

We believe that the concentrations of dissolved interstitial silica are controlled not by equilibrium processes between pore water and the associated solids, but rather by a dynamic balance between dissolution of solids (primarily amorphous biogenous silica, but not necessarily exclusively) and the diffusion of the dissolution products out of the pore waters into the overlying seawater. The dissolved silica comes from particles continuously stirred into the sediments by bioturbation, the principal process that redistributes reactive particles after they have fallen out of the water column onto the sea floor.

MODELLING INTERSTITIAL CONCENTRATIONS

The hypothesis that pore water is not in chemical equilibrium can be tested by calculations using diagenetic equations with parameters chosen as accurately as possible. It is necessary (although not sufficient) to our argument, that such calculations reproduce concentration profiles found in the sea floor. For these calculations we must include all the effects which have important influence on pore water composition. These include not only the aqueous phase processes, but also those factors influencing the distribution and reactions of the solid particles which participate.

Table I summarizes the relevant processes. These processes can be incorporated into a fairly realistic model of the sea-floor system by the following equations (Schink et al., 1975):

$$(\phi + K_D^*)(\partial c/\partial t) = \partial/\partial z[\phi D_c(\partial c/\partial z)] - v[\partial(\phi + K_D^*)c/\partial z] + K_B B(c_f - c)/c_f \qquad (1)$$

$$\partial B/\partial t = \partial/\partial z[D_B(\partial B/\partial z)] - v(\partial B/\partial z) - K_B B(c_f - c)/c_f \qquad (2)$$

The meaning of each of these terms is given more precisely in Schink et al. (1975). For convenience we will review them here:

ϕ = porosity (dimensionless)
K_D^* = effective adsorption coefficient of dissolved species on associated solids (dimensionless)
c = concentration of dissolved species in aqueous phase only (μmole cm^{-3})

c_f = concentration at saturation in the aqueous phase
c_0 = concentration in the seawater contacting the bottom
t = time (kyr)
z = distance from the sediment—water interface increasing downward (cm). Reference system is fixed to the interface.
D_c = diffusion coefficient in porous medium, D_m/θ^2 where D_m is aqueous diffusion coefficient in homogeneous phase (cm^2 kyr^{-1}) and θ is tortuosity (dimensionless)
v = apparent velocity of solid particles and pore water as viewed from the advancing sediment—water interface. Pore water and sediment move at the same velocity only when $\partial\theta/\partial z = 0$ (cm kyr^{-1})
K_B = first-order dissolution rate constant for substance dissolving to increase c (kyr^{-1})
B = concentration of dissolvable particulate material in bulk phase (solids + fluids) of the sediment (μmole cm^{-3})
D_B = rate of vertical mixing in the solid phases of sediments (cm^2 kyr^{-1})

Note that c and B are concentrations referenced to different volumes so that $-\partial B/\partial t = \partial\phi c/\partial t$ when transport processes are neglected. The boundary conditions for this model are simply: (1) that the input flux of particulates is equal to mixing redistribution plus advection (due to sediment accumulation) at the sea-floor boundary:

$$F_B = -D_B(\partial B/\partial z) + vB_0$$

(2) that the concentration in the aqueous phase at 50 cm above the bottom (c_w) be a measured value (in these first calculations $c_w = c_0$, the concentration at the interface); and (3) and (4) that there be no gradients at the bottom of the model.

To apply this model to real sediments we have first assumed that steady-state conditions apply. This is a reasonably valid assumption since the pore

TABLE I

PARAMETERS AFFECTING INTERSTITIAL SILICA CONC.

PRIMARY	1. INPUT FLUX OF BIOGENOUS SILICA
	2. BIOTURBATION RATE
	3. DISSOLUTION RATE
	4. BOTTOM WATER CONC.
SECONDARY	5. POROSITY & TORTUOSITY
	6. DIFFUSION COEFFICIENTS
	7. ADSORPTION COEFFICIENTS
	8. OTHER SOLID PHASE REACTION RATES
	9. ACCUMULATION RATES

water at the bottom of the model (1 m) would equilibrate with interface conditions in 100—300 yr.

The system has been simplified by treating porosity as a constant. Then at steady state the equations become:

$$0 = \phi D_c(\partial^2 c/\partial z^2) - v(\phi + K_D^*)\partial c/\partial z + K_B B(c_f - c)/c_f \quad (3)$$

$$0 = D_B(\partial^2 B/\partial z^2) - v(\partial B/\partial z) - K_B B(c_f - c)/c_f \quad (4)$$

This model requires that we specify ϕ, K_D^*, c_0, D_c, v, c_f, D_B, K_B and F_B as input parameters. Of these ϕ, c_0, D_c, v, and c_f are measurable and usually well known. We have tested this model with a broad range of input parameters, using the full span of observed values. Within the recognized scope of natural variation ϕ, D_c, K_D^*, and v have only minor (±25%) effect on interstitial concentration profiles. c_f has been measured several times with good agreement (e.g. Siever, 1962; Kato and Kitano, 1968; Hurd, 1972). Bioturbation rate (D_B) and dissolution rate (K_B) affect the system by their ratio unless the pore waters approach saturation; avoiding that condition we can combine these variables and treat them as one. Silica concentration in the overlying waters (c_w) varies with geographic regions, but usually is known. We have often measured it. This term too has only a minor effect on interstitial silica unless a boundary layer existed to impede diffusion and to raise the interfacial concentration c_0 above c_w. We shall return to that question. The input flux of biogenous particulate silica (F_B) turns out to be the most effective agent in determining interstitial silica concentration.

VALUES FOR THE PARAMETERS

Sediment mixing

Bioturbation rates are difficult, but not impossible, to quantify. Guinasso and Schink (1975) have re-examined the data of Glass (1969) on redistribution of microtektites and the data of Noshkin and Bowen (1973) on plutonium in deep-sea sediments. Treating the bioturbation process as equivalent to an eddy diffusion in an upper mixed layer of sediments (Fig.3) they obtained the time-dependent solution for redistribution of a tracer pulse or any other input function. The intensity of mixing can be described by the dimensionless parameter $G = D_B/Lv$ where L is the thickness of the mixing layer (cm) and other terms are as previously identified: D_B the bioturbation diffusion coefficient and v the sediment accumulation rate.

Some solutions to this equation are shown in Fig.4. For low values of G (e.g. $G = 0.03$) bioturbation is relatively less intense and burial of the tracer pulse proceeds as the tracer spreads slightly into almost a Gaussian distribution. Concurrently, bioturbation plus sediment accumulation move the concentration maximum downward only slightly faster than it would be moved by sediment accumulation acting alone.

Where G is large (e.g. $G = 3$) the relatively intense bioturbation homogenizes the mixed layer and the ultimate distribution of tracer shows a

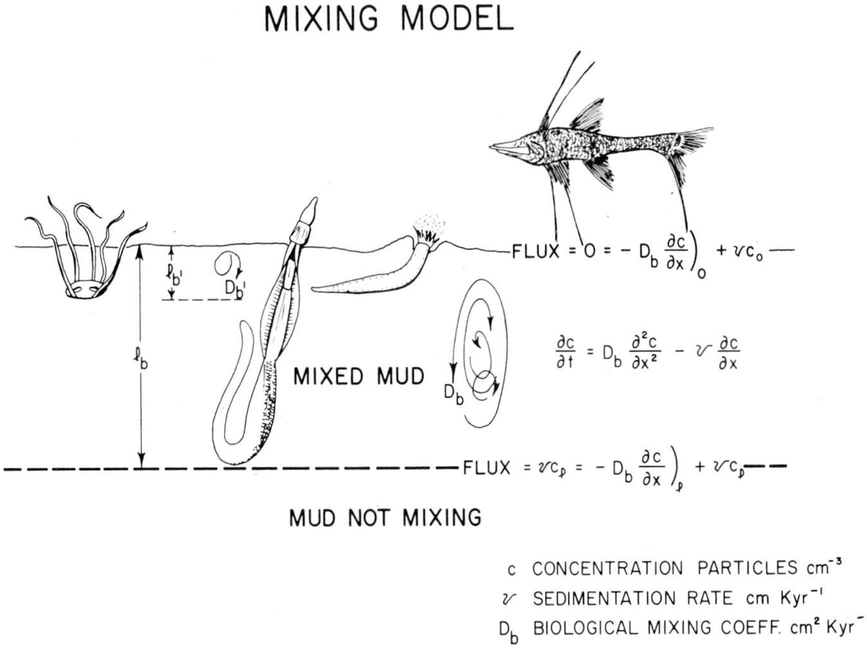

Fig.3. Dramatization of the mixing model. Organisms not to scale. D_b' and l_b' denote the possibility of more than one mixing rate at more than one mixing length.

sawtooth character with the concentration maximum displaced by one mixing length below the level that it would have were sediment accumulation the only process operating.

Fig.5 shows a simple practical method for estimating the bioturbation parameters based on the observed redistribution of an input pulse of stable tracer. First, determine the standard deviation and skewness of the final distribution after biological mixing is terminated by burial. From the observed skewness use Fig.5 to estimate the value of G on this sample. From the G value thus obtained use the other curve to estimate σ/L.

Fig.4. Moving picture of concentration profiles at progressive times for several identical impulse sources subjected to different mixing conditions. The mixed layer is indicated by stippling. Time t^* has been normalized to the period required for sediments of one mixed layer thickness to accumulate; i.e., after one such interval, $t^* = 1$. Depth is normalized to the mixing layer thickness L; i.e., depth is 1 at the bottom of the mixed layer. The vertical series represent the profiles of the tracer distribution at progressive times increasing from top to bottom. Each vertical series represents a different mixing condition as defined by the mixing parameter $G = D/Lv$. The vertical series on the left shows the least mixing. In this series, little mixing takes place before the impulse is buried below the zone of mixing. The central series are intermediate. On the right the mixing rate is great enough to homogenize the mixed layer in a time short compared to the residence time in the mixed layer. See Guinasso and Schink (1975) for derivation of these curves.

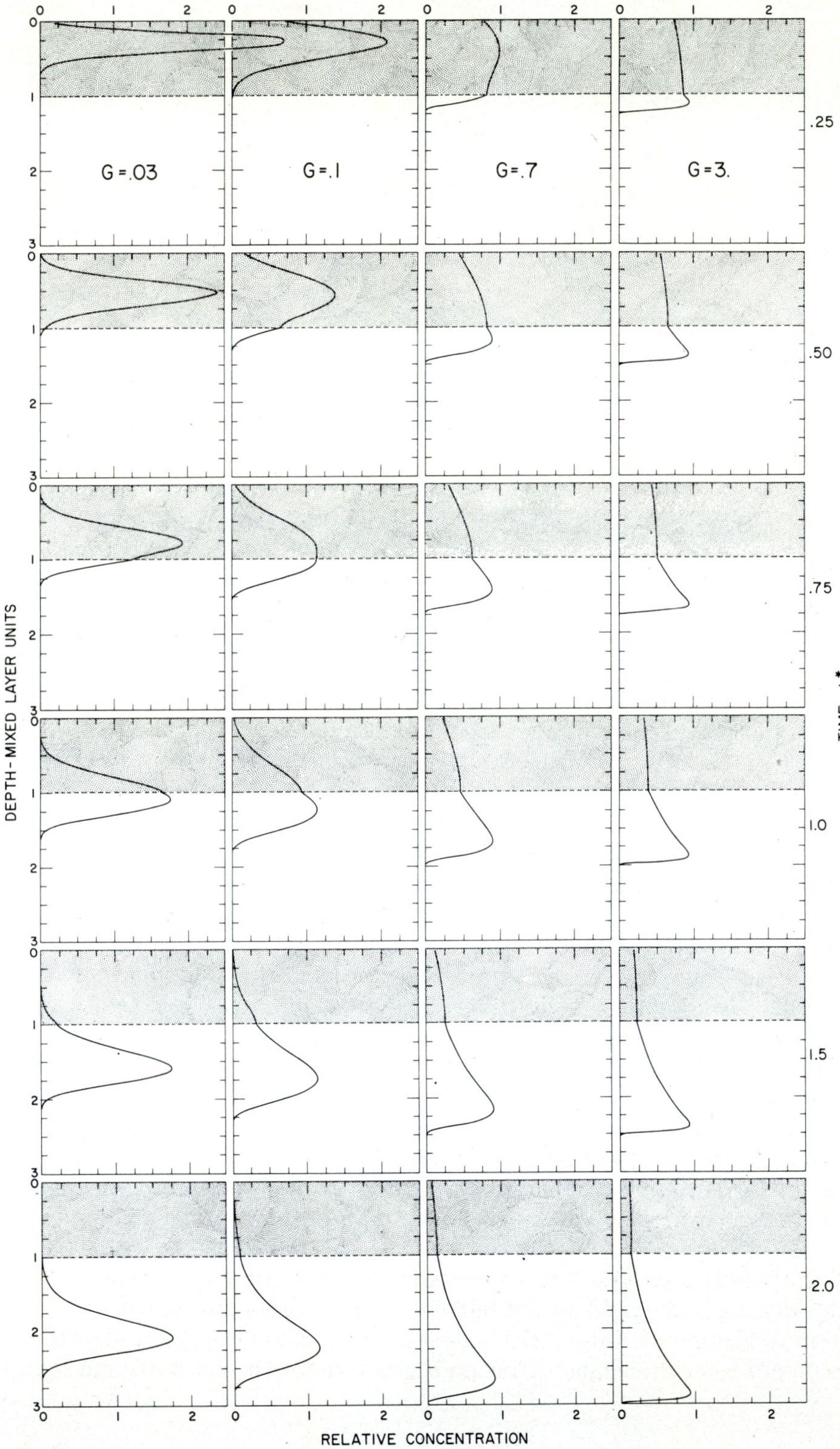
G = .03
G = .1
G = .7
G = 3.
.25
.50
.75
1.0
1.5
2.0
TIME t*
DEPTH - MIXED LAYER UNITS
RELATIVE CONCENTRATION
0
1
2
3

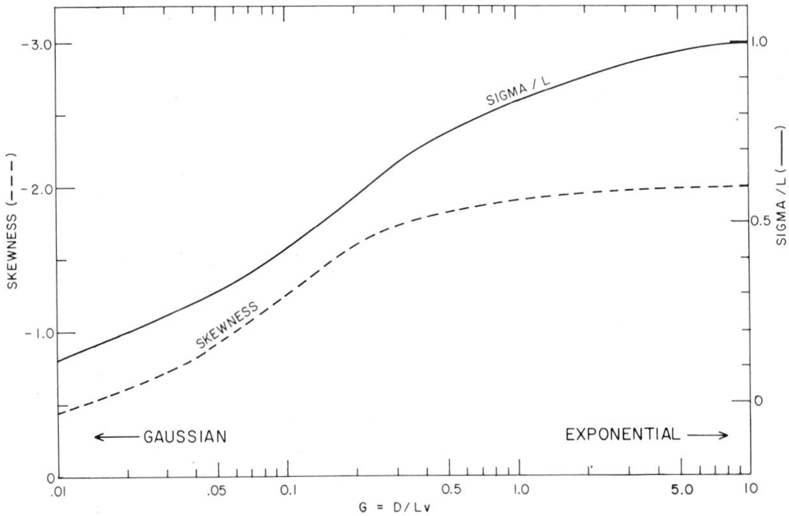

Fig.5. Standard deviation and skewness as functions of G from Guinasso and Schink (1975). The standard deviation, normalized to the mixing depth, is plotted as σ/L. Note that the two vertical scales are offset. Skewness ranges from zero (Gaussian) to −2 (exponential).

Knowing σ allows one to estimate L. Knowing G and L allows us to evaluate D_B if we can determine sediment accumulation rates by other means. Knowing the time of the Australasian tektite event, plus the depth of burial, provides one such means.

Interpreting the distribution of a contemporary tracer is even simpler. We assume that sediment accumulation is nil in the 10—20 years of plutonium accumulation at the sea floor. Then redistribution is given by $\sigma = (2D_B t)^{1/2}$. Only a lower-limit for mixing length can be determined from this data. Table II gives various mixing rates as determined by Guinasso and Schink (1975). The values from plutonium data are much more rapid than those determined from tektite distributions. This is because the stable tracers are most sensitive to those mixing processes with the longest mixing length even though such processes may be the result of very rare events and thus have low values of D_B.

For a dissolving particle, the more rapid processes operating over shorter lengths have greater effect on the redistribution. Accordingly we have chosen values in the range 100—300 cm^2 kyr^{-1} for our biological mixing coefficients. These "rapid" values are still much slower than the slow process of aqueous molecular diffusion.

Bioturbation is the result of an occasional worm eating its way down to greater depths in the mud and of burrowers and worms turning over the uppermost sediments. Interstitial waters are refreshed or exchanged in this process, but since molecular diffusion is much faster (in the statistical sense)

TABLE II

Biological mixing rates

A. Microtektite data[*1]

Core	Depth of maximum conc. (cm)	Mean depth (cm)	σ (cm)	Skewness	Sedimentation rate (cm kyr^{-1})	G	σ/L	L (cm)	D ($cm^2\,kyr^{-1}$)
RC9-143	349	343	9.3	−1.32	0.51	0.13	0.43	21.9	1.4
V19-153	510	510	7.2	−0.63	0.75	0.02	0.21	34.0	0.58
V19-297	541	526	17.2	−2.07	0.74	>10.00	1.00	17.2	>130
V20-138	330	322	9.7	−0.68	0.46	0.03	0.23	42.0	0.52
V16-70	155	154	9.7	−1.38	0.22	0.15	0.44	21.9	0.75
V16-76	1043	1037	9.3	−1.22	1.51	0.09	0.41	22.6	3.2
RC9-58	254	245	20.2	−4.92	~1.00	>10.00	1.00	20.2	>200
RC9-137	414	407	8.1	−0.50	0.64	0.02	0.17	48.3	0.49

B. Plutonium data[*2]

Core	Latitude	Longitude	Depth (m)	Month/year collected	Maximum Pu penetration (cm)	σ[*3] (cm)	Mixing rate[*4], D ($cm^2\,kyr^{-1}$)
AII-49-2	41°21′N	08°41′E	1000	V-69	12	2.7	380
AII-49-3	39°02′N	42°36′W	4810	VII-69	8	2.1	220
AII-59-3	21°54′N	18°17′W	1410	VI-70	9	2.2	250
AII-60-1	29°59′S	04°55′E	4920	V-71	4	1.7	140
AII-60-3	09°35′S	12°20′E	1345	VI-71	6	1.4	100

[*1] B.P. Glass (personal communication, 1973).

[*2] Noshkin and Bowen (1973).

[*3] $\sigma = \Sigma z^2\alpha \;/\; \Sigma\alpha$ where z is the depth of the center of the interval over which the measurement was made and α is the measured plutonium activity.

[*4] The value D was calculated assuming 10-year mixing time (T = 10 yr) from the formula $D = \sigma^2/2T$.

aqueous gradients are quickly re-established and the effect on aqueous fluxes and concentrations is small. Effects of bioturbation on aqueous gradients cannot be neglected in regions where bioturbation rates are 10 or more times faster than we have estimated. Such a situation must occur commonly on continental shelves in fertile regions. There "ventilation" of pore waters by bioturbation can be an important effect. In the abyss ventilation does not appear significant.

Particle dissolution

Dissolution rates are, surprisingly, even harder to quantify. Such rates have been measured by a number of investigators (Kamatami, 1969, 1971; Hurd, 1972; Johnson, 1975). But the conditions of the experiments have always been such that the results were inappropriate to the actual environmental conditions of dissolution. Both Hurd and Johnson cleaned their samples with peroxide and acid. Kamatami dissolved his samples at elevated temperatures. In each case the natural dissolution rate was substantially altered. What data we have suggest that siliceous frustules should last days, or at most a few years on the sea floor. These experimental data are contradicted by natural observations. Siliceous fossils many millions of years old can be found in undersaturated pore waters. Radiolaria and even diatoms are commonly buried intact with the sediments. The mechanism protecting these survivors is not well understood. Lewin (1961) showed that inorganic coating could protect siliceous tests and retard or prevent dissolution. Iron, manganese and aluminum might serve to protect the tests. Lewin's data are hard to apply to the sea-floor situation. Clearly most silica tests dissolve, some do not. Two hypotheses might be offered: (1) there are different types of silica tests reaching the sea-floor — some soluble, some insoluble; (2) there is a race between the dissolution process and the formation of a protective film — possibly complicated by the decomposition of an initial organic film — accompanied by replacement with some more permanent protection.

Somehow these processes operate to dissolve most of the silica falling to the sea-floor, but permit a tiny fraction to persist and be buried with the other sediments. Curiously, many sediments with high opal content do not have saturated pore waters. We do not know how to distinguish these insoluble particles visually, but mathematically we simply exclude them from our definition of B and F_B.

Hurd (1972) has described dissolution of silica from siliceous frustules in terms of a rate constant (mg cm^{-2} sec^{-1}) which involves consideration of surface area. This is probably the correct approach, but it leaves some uncertainty as to whether the surface area, as measured by nitrogen adsorption, is the appropriate surface area for dissolution.

Because siliceous tests are so complex we have assumed that specific surface (i.e. surface area per unit mass) will remain essentially constant as the silica dissolves. This assumption allows us to relate dissolution to the mass of dissolvable silica without measuring the surface area of each sample. This is a convenient approximation. Since the actual in situ dissolution rate is extremely uncertain, any argument over the validity of this simplification must, for now, be academic.

The dissolution rates we have chosen are based on unpublished data (Johnson, personal communication, 1974) involving dissolution of uncleaned radiolaria from the sea floor. Values center on 90 kyr^{-1}.

From these considerations the conclusion emerges that no one has an accurate quantitative description of dissolution rates or bioturbation rates. We need more precise measurements and we need measurements appropriate to the specific location of interest. But these shortcomings should not prevent further considerations of the system. By assuming rates of bioturbation and of dissolution ranging over two or three orders of magnitude we can calculate resulting interstitial silica profiles and compare them with observation. Such comparisons help to sharpen our thinking and focus measurements on the most likely range of values. Calculations of this sort might also show that our approach to the problem is all wrong. In this instance they suggest that we are headed in the right direction.

Input rates

Values for the input flux of biogenic silica have never really been measured, but they are much more tightly constrained (at least on the average) than are bioturbation or dissolution rates. We know, for example, that somewhere between 2 and 4 atoms of carbon are fixed for each atom of silicon incorporated into organisms. Carbon productivity estimates range from 100—500 mg m^{-2} day^{-1}. This then suggests silica fixation at a rate of 75—150 μmole cm^{-2} yr^{-1}. Berger (1970) and Lisitzin (1972) using similar arguments suggested values of ~12—140 μmole cm^{-2} yr^{-1}. Probably less than 10% of this reaches the sea-floor (Calvert, 1968).

Deep-ocean dissolved silica increases with the age of the water. Young North Atlantic Deep Water has about 25 μM SiO_2 in solution. In the North Pacific the deep water, roughly 1500 years older, has a silica content of about 160 μM. Although some of this increase is due to dissolution of silica particles as they sink, as an extreme case we might consider the increase to be entirely from dissolution at the bottom, followed by mixing upward through 3000 m of water. This increase of dissolved silica in deep water then suggests that the flux from the deep sea-floor cannot average more than about 30 μmole cm^{-2} yr^{-1}.

Fanning and Pilson (1974) measured the silica flux from some Atlantic cores. They found from 3 to 18 μmole cm^{-2} yr^{-1}. Their values seem reasonable in comparison with previous considerations, although higher values would be expected under more fertile regions. Accordingly, the input flux of soluble particulate silica seems rather well defined within 3—30 μmole cm^{-2} yr^{-1} with the low and the high values representing fluxes from under barren waters or from under fertile waters respectively. Sediment trap experiments or some other observational approaches are needed to verify these values, but we feel reasonably confident that they are close.

While considering these values it is useful to compare the flux of biogenous silica which is buried with accumulating sediments. A sediment composed of 30% opal, depositing at 3 cm kyr^{-1}, with porosity of 0.9 and average solids density of 2.6 g cm^{-3} represents a silica removal of 4 μmole cm^{-2} yr^{-1}. Apparently siliceous ooze even when accumulating fairly rapidly can retain only a minor fraction of all the silica delivered to the sea-floor.

Another useful comparison is the ocean-wide average silica removal to the sea-floor. River input of dissolved silica provides an annual contribution of 2 μmole cm^{-2} yr^{-1} if deposition were equal to input, and if the deposits were distributed evenly over the entire ocean floor.

Other parameters

In order to make the model calculations we have fixed most of the other parameters at some reasonable value. Values selected were:

$\phi = 0.8$
$D_c = 4 \cdot 10^{-6}$ cm^2 sec^{-1}
$= 126{,}000$ cm^2 kyr^{-1}
$v = 2$ cm kyr^{-1}
$K_D{}^* = 0$

CALCULATION

Eqs. 3 and 4 were solved by numerical methods. Details of the calculation methods and the techniques used to verify the results are given in Schink et al. (1975). A report describing the computer program used to evaluate the solutions is available on microfilm from the American Geophysical Union.

Results

Some results of calculations are shown in Figs.6—9. Fig.6 shows the strong effect that variations in input rate can have on the interstitial silica profile. These calculations support suggestions made in Schink et al. (1974). Fig.7 shows the effect of variations in the ratio of bioturbation rate to dissolution rate. Rapid bioturbation causes the reactant particle to be displaced downward where its effect on pore water composition can be much greater. Slower dissolution permits more time for this downward mixing and thus enhances the effect. It may seem surprising that slower dissolution leads to higher pore water concentrations. This effect depends on the assumption that pore waters do not saturate, that the protective mechanism does not set up more effectively on slow dissolvers, and that all soluble particles dissolve sooner or later. Under these conditions, slower solution leads to higher concentrations in solution.

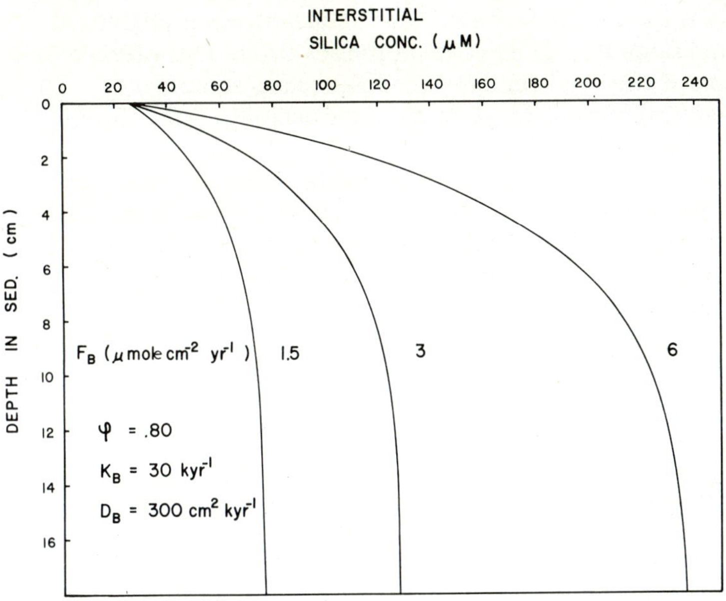

Fig.6. Profiles of interstitial silica versus depth in sediment calculated for various input rates of dissolvable particulate silica.

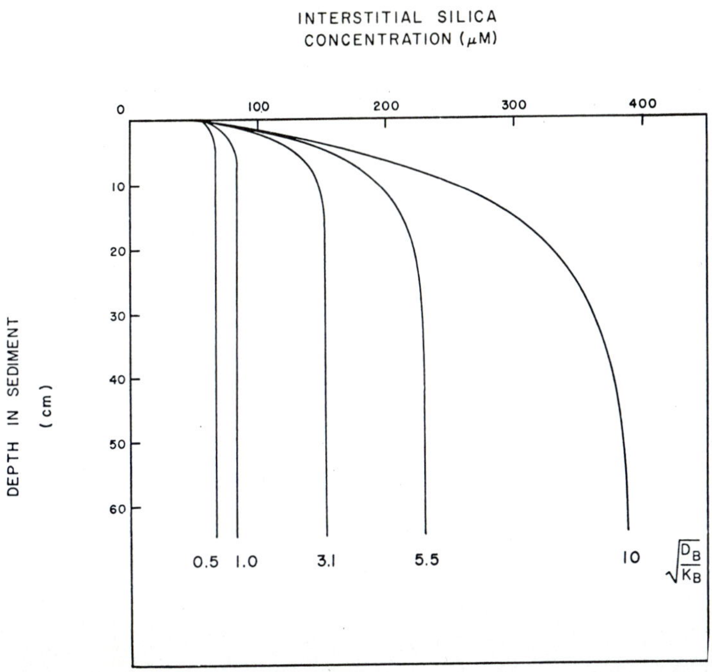

Fig.7. Profiles of interstitial silica versus depth calculated for a variety of values of the ratio D_B/K_B. All profiles have the same slope at the interface.

Also noteworthy is the fact that the different concentration profiles in Fig.7 all produce the same flux of silica from the sea-floor. This offers a dramatic illustration of the need for careful, close-spaced sampling of undisturbed sediment samples if we are to produce accurate estimates of the silica flux from the sea-floor.

Fig.8 shows the distribution of soluble particulate biogenous silica as calculated for conditions identical to those in Fig.6. This shows that changes almost unobservable in the solid phases can support changes easily measured in pore water composition. Recall that B is defined as the concentration of particulate silica *that will dissolve.* Sediments often contain 0.5—2% amorphous silica that will not dissolve. We don't know how to distinguish the dissolvable from the undissolvable visually, so there are often no detectable differences in the sum of the two components as soluble particles disappear. However, Schrader (1972) observed the removal of more soluble siliceous forms over the appropriate depths in the core.

Fig.8 offers an interesting contrast in silica fluxes. 0.5% SiO_2 in these sediments, if buried with the sediment, would represent a loss of less than 0.1 μmole cm^{-2} yr^{-1}. Whereas the dissolving silica, constituting 0.2—0.8% of the solids at maximum, generates fluxes in the aqueous phase that are 15—17 times greater.

Fig.9 shows the effect of varying the bottom water concentration. Normally this concentration is fixed by oceanographic factors, but varying

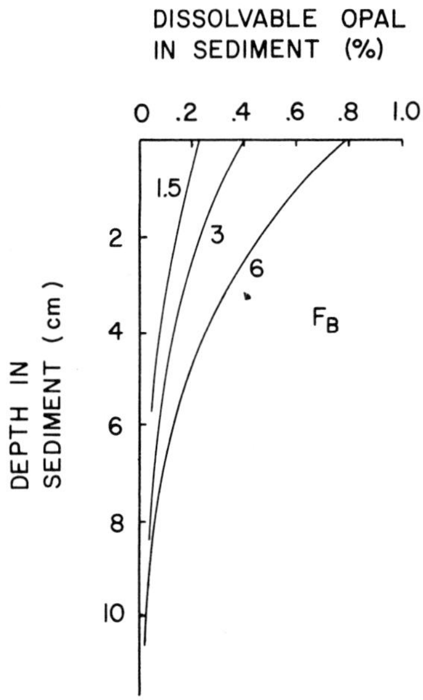

Fig.8. Profiles of *dissolvable* particulate silica in sediments for conditions identical to those in Fig.7. Note the scale for weight percent is from 0 to 1% of the total solids in the sediments.

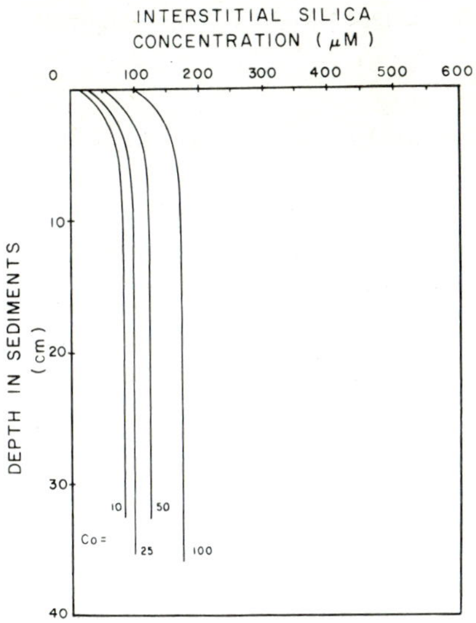

Fig.9. Profiles of interstitial silica versus depth in sediment for various values of c_0. Other parameters held constant: F_B = 3 μmole cm^{-2} yr^{-1}; D_B = 450; K_B = 90.

conditions over time might cause shifts at a given location. In all oceans the concentration of silica in near-bottom waters is so far below saturation that this effect seems most unlikely as an explanation for variations in silica preservation. The only possible exception might be if dissolution of opal were not the reaction that saturates pore water and stops dissolution. If some other reaction were the effective one, and the saturation concentration were closer to that of overlying waters, then variations in the dissolved silica content of near-bottom waters could play a more significant role.

IMPROVING THE MODEL

In an attempt to improve upon the modelling described in Schink et al. (1975) we have introduced some additional features to the calculation. We have added a variable mixing coefficient, we have included sediment accumulation, and we have considered the non-turbulent boundary layer as a further impedance to the diffusive flux.

We have calculated interstitial concentrations and biogenous silica profiles for various patterns of bioturbation. We let D_B be constant with depth and imposed no "mixing depth" cut-off; we let D_B decrease linearly from the interface to a zero value at mixing depth (10 cm); we let D_B decrease exponentially with a scale length equal to the "mixing depth" (10 cm). Among all of these cases we found negligible differences in the biogenous silica profiles and less than 20% differences in the dissolved silica profiles when the final dissolved silica concentration was low and the dissolvable silica disappeared at fairly shallow depths.

When the pore water saturates, the mixing function should again become unimportant, with the balance between aqueous diffusion and dissolution establishing the profiles. However, the nature of the bioturbation mixing function will have effect on profiles approaching or just achieving saturation; we have not explored these conditions thoroughly.

Morse (1974) suggested that the non-turbulent boundary layer at the sea-floor could substantially reduce chemical fluxes across this zone. Morse's arguments were based on the assumption that some process fixes the concentration in interstitial water so that impedance variations change the flux. If the interstitial concentrations are the result of competing effects, as we propose, then added impedance would simply cause the concentrations to rise; fluxes would be little altered.

To examine this problem the silica model calculations were extended into the water column, where diffusivity varies with distance from the boundary. In this zone:

$$\frac{\partial c}{\partial t} = \frac{\partial}{\partial z} D(z)\, \partial c/\partial z \qquad (5)$$

where $D(z)$ is the coefficient of diffusivity given as a function of height above the bottom. Descending into the boundary region, transport mechanisms are transformed from those dominated by turbulent processes to those dominated by molecular motion. The transformation is not a simple one. Turbulence is created by the drag of the sea-floor on water driven by deep-sea currents. This turbulence mixes the waters overlying the sea bed, but the turbulent eddies are damped and become progressively less effective as we approach the interface. The system is further complicated by periodic variations in velocity of deep-sea currents, so the regime is constantly changing its character. Wimbush and Munk (1970) distinguish two regions in the vicinity of the sea-floor: (1) just above the interface is the viscous sublayer; (2) transitional to the sea above is the logarithmic layer. Morse (1974) estimated chemical fluxes by assuming that diffusion of a chemical tracer takes place only by molecular diffusion in the viscous sublayer. However Wimbush (personal communication, 1976) has pointed out that this is not the correct treatment for mass fluxes. In a region where momentum is being transferred mostly by viscosity, heat and chemical tracers can still be transferred primarily by the turbulent motions. Only when turbulent transports of heat and mass become less than their associated molecular transports can one say that a "boundary layer" exists, where heat transfer is dominated by thermal diffusivity and mass transfer by molecular diffusivity. Wimbush (1976) discusses these concepts in more detail.

The viscous sublayer is distinguished from the zone above in that turbulent transfer of momentum has become less than molecular transfer of momentum. This does not require that turbulent fluxes are non-existent, but only that they are much less effective. In seawater (at 2°C) the thermal diffusivity ($1.4 \cdot 10^{-3}$ cm^2 sec^{-1}) and chemical diffusivity ($5 \cdot 10^{-6}$ cm^2 sec^{-1}) are much less than the kinematic viscosity ($1.7 \cdot 10^{-2}$ cm^2 sec^{-1}). Although momentum transport in the viscous sublayer is dominated by viscosity, turbulent transport of heat and chemical tracers can still be accomplished by turbulent processes. Viscosity, then, becomes the major transmitter of momentum at a boundary layer thickness far greater than the thickness of the layer in which molecular diffusion processes dominate turbulent diffusion. The "diffusion sublayer" is therefore much thinner than the viscous sublayer.

To model the effects of the non-turbulent boundary zone we used the molecular diffusion coefficient to describe $D(z)$ in the diffusion sublayer. The layer thickness can be estimated from experimental data. Hinze (1959) describes a transfer coefficient C_c appropriate to this type of situation as:

$$C_c = F/U_0 \, \Delta c \tag{6}$$

where F is the flux of mass across the zone (μmole cm^{-2} sec^{-1}), Δc is the concentration change across the zone (μmole cm^{-3}) and U_0 is the free stream velocity in the water above this zone (cm sec^{-1}). An analogous coefficient, C_h, can also be defined for heat.

In order to estimate the thickness of the diffusion sublayer, we assume the chemical flux across this layer can be described by Fick's first law:

$$F = D_m \, \Delta c/\delta \tag{7}$$

where δ is the thickness (cm) of the diffusion sublayer. Combining eqs. 6 and 7 we find:

$$\delta = D_m/U_0 C_c \tag{8}$$

Deissler (1954) measured and compiled data for C_c as a function of Schmidt number (Sc) and for C_h as a function of Prandtl number (Pr) for various Reynolds numbers. The data are summarized in Fig.10. The Schmidt number is the ratio of viscosity to molecular diffusivity; the Prandtl number is the analogous coefficient for thermal diffusivity. For dissolved silica in seawater (S = 35%, at 2°C) the numerical value of Sc is 3500. For seawater at 2°C, Pr is about 12. The figure shows C_c or C_h plotted against Sc or Pr. The data for Sc or Pr <100 are from heat flux measurements while the data for Sc or Pr >100 are from chemical diffusion measurements. The data show that C_c or C_h is roughly proportional to Sc or Pr to the −4/3 power, but also depend on the Reynolds number. For most near-bottom conditions the Reynolds number will be in the range of $1 \cdot 10^4$ to $5 \cdot 10^4$. Variations in Reynolds number of this magnitude shift the relevant C_c at Sc = 3500 values only from $1.3 \cdot 10^{-5}$ to $1.7 \cdot 10^{-5}$.

Using the median value for C_c of $1.5 \cdot 10^{-5}$ and $5 \cdot 10^{-6}$ cm^2 sec^{-1} for D_m we find:

$$\delta = \frac{1}{3U_0} \tag{9}$$

Free current velocity U_0 depends slightly upon the height of measurement but the actual choice has little effect on the outcome. For near-bottom

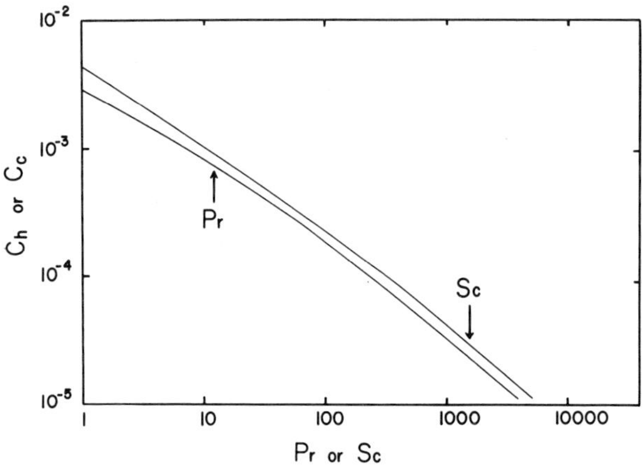

Fig.10. A plot of the mass transfer coefficient C_c against Schmidt number (Sc) or the heat transfer coefficient C_h plotted against the Prandtl number (Pr). The curves are based on experimental data compiled by Deissler (1954). The arrows represent Pr for seawater at 2°C and Sc for silica in seawater at 2°C. The upper line represents C_c or C_h for a Reynolds number (Re) of 10^4; the lower line is for Re equal to $5 \cdot 10^4$. This represents the expected range of Reynolds numbers for flow regimes near the sea floor.

currents in the range 1—10 cm/sec we would expect boundary thicknesses of 3 to 0.3 mm.

Above this zone, in the logarithmic layer the vertical eddy diffusivity increases with distance above the bottom according to:

$$D(z) = ku_*z \tag{10}$$

where k is the von Kármán constant equal to 0.4 and u_* values have been measured by Wimbush to fall in the range 0.03—0.3 cm sec^{-1}. From eq. 10 we establish $D(z)$ and use this to calculate concentration gradients above the boundary zone. A simple calculation shows these gradients quickly disappear. Our analysis shows that the diffusion boundary layer is too thin to have significant effect on the silica flux or on the interstitial silica profiles.

We have described here two attempts to improve the model calculations described in Schink et al. (1975). The complex model is perhaps more realistic, but the results suggest no new insights. The conclusions previously reached appear relatively insensitive to the nature of the bioturbation function, or to the "blanketing" effect of a stagnant boundary layer.

CONCLUSIONS

These calculations support the contention that interstitial silica profiles represent dynamic balance and not equilibrium conditions. Silica dissolution is influenced by the departure from saturation, but a protective mechanism — not understood — also plays an important role. Since most interstitial waters are well undersaturated, variations in the "corrosive effects" of overlying bottom water do not play a large role in controlling dissolution of siliceous microfossils; temperature and pressure effects on the solubility of SiO_2 are far too mild to exert any sort of strong influence; silica differs significantly from $CaCO_3$ in this respect. Stagnant boundary layers do not play a significant role in regulating the flux or the concentration of dissolved interstitial silica.

The greatest influence on interstitial silica concentration is exerted by the input flux. Secondarily, the bioturbation rate, dissolution rate, the "mysterious" protective mechanism and the overlying water concentration all can alter the balance in the sediments.

Any change in bioturbation rates or in the flux of dissolvable particles to the sediment will tend to shift the balance concentration in the dissolved silica profile. Such shifts would first appear at the surface and would propagate downward in the pore water, losing fine structure as the changes propagated. The rate of transmission of such changes is far more rapid than the sediment accumulation rates. Accordingly we might note that the "age" of the pore waters in the upper meters is far less than the age of the associated sediments, and that these waters contain a record — albeit with poor resolution and still hard to read. This record could tell us something about the balance of factors that established pore water concentrations over the past hundred years in this region, if we learn to read it.

ACKNOWLEDGEMENTS

The authors are indebted to Kent Fanning for his part in developing the dissolved silica models and to Mark Wimbush for setting us straight on some aspects of boundary layer theory. Thanks are also due to A.D. Kirwan Jr. for his helpful comments and to Bruce Heezen for his tolerance in accepting our late submissions. This work was supported by the Office of Naval Research Contract No. N00014-75-C-0537.

REFERENCES

Berger, W.H., 1968. Planktonic Foraminifera: selective solution and paleoclimatic interpretation. Deep-Sea Res., 15: 31—43.

Berger, W.H., 1970. Biogenous deep-sea sediments: fractionation by deep-sea circulation. Geol. Soc. Am. Bull., 81: 1385—1402.

Calvert, S.E., 1968. Silica balance in the ocean and diagenesis. Nature, 219: 919—920.

Deissler, R.G., 1954. Analysis of turbulent heat transfer, mass transfer, and friction in smooth tubes at high Prandtl and Schmidt numbers. Nat. Advis. Comm. Aeronaut., Tech. Notes, 3145: 53 pp.

Fanning, K.A. and Pilson, M.E.Q., 1974. Diffusion of dissolved silica out of deep-sea sediments. J. Geophys. Res., 79: 1293—1297.

Glass, B.P., 1969. Reworking of deep-sea sediments as indicated by the vertical dispersion of the Australasian and Ivory Coast microtektite horizons. Earth Planet. Sci. Lett., 6: 409—415.

Guinasso Jr. N.L. and Schink, D.R., 1975. Quantitative estimates of biological mixing rates in abyssal sediments. J. Geophys. Res., 80: 3032—3043.

Hinze, J.O., 1959. Turbulence. McGraw Hill, New York, N.Y., 586 pp.
Hurd, D.C., 1972. Factors affecting solution rate of biogenic opal in seawater. Earth Planet. Sci. Lett., 15: 411—417.
Johnson, T.C., 1975. The Dissolution of Siliceous Microfossils in Deep-Sea Sediments. Thesis, Univ. Calif., San Diego, Calif.
Jones, M.M. and Pytkowicz, R.M., 1973. Solubility of silica in sea water at high pressures. Bull. Soc. R. Sci. Liege, 42: 118—120.
Kamatami, A., 1969. Regeneration of inorganic nutrients from diatom decomposition. J. Oceanogr. Soc. Jpn., 25: 63—74.
Kamatami, A., 1971. Physical and chemical characteristics of biogenous silica. Mar. Biol., 8: 89—95.
Kato, K. and Kitano, H., 1968. Solubility and dissolution rate of amorphous silica in distilled and sea water at 20°C. J. Oceanogr. Soc. Jpn., 24: 147—152.
Lewin, J.C., 1961. The dissolution of silica from diatom walls. Geochim. Cosmochim. Acta, 21: 182—198.
Lisitzin, A.P., 1972. Sedimentation in the World Ocean. Soc. Econ. Paleontol. Mineral., Spec. Publ., 17: 218 pp.
Morse, J.W., 1974. Calculation of diffusive fluxes across the sediment—water interface. J. Geophys. Res., 79: 5045—5048.
Noshkin, V.E. and Bowen, V.T., 1973. Concentrations and distributions of long-lived fallout radionuclides in open ocean sediments. In: Radioactive Contamination of the Marine Environment. International Atomic Energy Agency, Vienna, pp.671—686.
Schink, D.R., Fanning, K.A. and Pilson, M.E.Q., 1974. Dissolved silica in the upper pore-waters of the Atlantic Ocean floor. J. Geophys. Res., 79: 2243—2250.
Schink, D.R., Guinasso Jr. N.L. and Fanning, K.A., 1975. Processes affecting the concentration of silica at the sediment-water interface of the Atlantic Ocean. J. Geophys. Res., 80: 3013—3031.
Schrader, H.J., 1972. Kieselsaure-Skelette in Sedimenten des ibero-marokkanischen Kontinentalrandes und angrenzender Tiefsee-Ebenen, "Meteor"-Forschungsergeb., 8: 10—36.
Siever, R., 1962. Silica solubility, 0—200°C and the diagenesis of siliceous sediments. J. Geol., 70: 127—150.
Willey, J.D., 1974. The effect of pressure on the solubility of amorphous silica in seawater at 0°C. Mar. Chem., 2: 239—250.
Wimbush, A.H.M.H. and Munk, W., 1970. The benthic boundary layer. In: A.E. Maxwell (Editor), The Sea, 4 (Part 1). Wiley, New York, N.Y., pp.731—758.
Wimbush, M., 1976. The physics of the benthic boundary layer. In: I.N. McCave (Editor), The Benthic Boundary Layer. Plenum, New York. N.Y., pp. 3—10.

Marine Geology, 23 (1977) 155—172

SUSPENDED PARTICULATE LOADS AND TRANSPORTS IN THE NEPHELOID LAYER OF THE ABYSSAL ATLANTIC OCEAN*

PIERRE E. BISCAYE and STEPHEN L. EITTREIM

Lamont-Doherty Geological Observatory of Columbia University, Palisades, N.Y. 10964 (U.S.A.)
U.S. Geological Survey, Menlo Park, Calif. 94025 (U.S.A.)

(Received July 1, 1976)

ABSTRACT

Biscaye, P.E. and Eittreim, S.L., 1977. Suspended particulate loads and transports in the nepheloid layer of the abyssal Atlantic Ocean. Mar. Geol., 23: 155—172.

Vertical profiles of light scattering from over 1000 L-DGO nephelometer stations in the Atlantic Ocean have been used to calculate mass concentrations of suspended particles based on a calibration from the western North American Basin. From these data are plotted the distributions of particulate concentrations at clear water and in the more turbid near-bottom water. Clear water is the broad minimum in concentration and light scattering that occurs at varying mid-depths in the water column. Concentrations at clear water are as much as one-to-two orders of magnitude lower than those in surface water but still reflect a similar geographic distribution: relatively higher concentrations at ocean margins, especially underneath upwelling areas, and the lowest concentrations underneath central gyre areas. These distributions within the clear water reflect surface-water biogenic productivity, lateral injection of particles from shelf areas and surface circulation patterns and require that the combination of downward vertical and horizontal transport processes of particles retain this pattern throughout the upper water column.

Below clear water, the distribution of standing crops of suspended particulate concentrations in the lower water column are presented. The integration of mass of all particles per unit area (gross particulate standing crop) reflects a relative distribution similar to that at the surface and at clear water levels, superimposed on which is the strong imprint of boundary currents along the western margins of the Atlantic. Reducing the gross particulate standing crop by the integral of the concentration of clear water yields a net particulate standing crop. The distribution of this reflects primarily the interaction of circulating abyssal waters with the ocean bottom, i.e. a strong nepheloid layer which is coincident with western boundary currents and which diminishes in intensity equatorward.

The resuspended particulate loads in the nepheloid layer of the basins west of the Mid-Atlantic Ridge, resulting from interaction of abyssal currents with the bottom, range from $\sim 2 \cdot 10^6$ tons in the equatorial Guyana Basin to $\sim 50 \cdot 10^6$ tons in the North American Basin. The total resuspended particulate load in the western basins ($111 \cdot 10^6$ tons) is almost an order of magnitude greater than that in the basins east of the Mid-Atlantic Ridge ($13 \cdot 10^6$ tons). The net northward flux of resuspended particles carried

---

* Contribution No. 2398 of the Lamont-Doherty Geological Observatory.

in the AABW drops from $\sim 8 \cdot 10^6$ tons/year between the southern and northern ends of the Brazil Basin and remains $\sim 1 \cdot 10^6$ tons/year across the Guyana Basin.

INTRODUCTION

The importance of suspended particulate matter in oceanic processes is only recently becoming appreciated. As extractors from, transporters through and sources to the water column of many major and minor elements, suspended particulate matter is responsible for maintaining most oceanic chemical concentration gradients. Particulates reflect the biological, geochemical, atmospheric and geological processes of yesterday and they are the deep-sea sediments of tomorrow.

The most extensive body of data on the distribution of suspended particulates in the world oceans is the more than 4,000 nephelometer profiles collected by the Lamont-Doherty Geological Observatory (L-DGO) in the twelve years since Edward Thorndike built the first L-DGO-Thorndike photographic nephelometer and Maurice Ewing used it to obtain vertical profiles of light scattered by suspended particles (Ewing and Thorndike, 1965). Numerous papers since that time have used the data obtained with this instrument in regional studies. What has emerged is a somewhat patchy picture of intense abyssal circulation reflected in the layer of turbid water called the nepheloid layer adjacent to much of the ocean bottom (for example see Eittreim et al., 1969, 1972; Ewing and Connary, 1970; Ewing et al., 1971; LePichon et al., 1971; Eittreim and Ewing, 1972). More recently, because of improvements in instrumentation and in data coverage, reports of nephelometer data on oceanic scales have begun to be published (Eittreim and Ewing, 1974; Kolla et al., 1976; Eittreim et al., 1976).

The data reported in these papers have been primarily in units of light scattering which conveys internally consistent information on the relative turbidity of the water in different parts of the water column, in different locations and even at different times. But these units are not meaningful or useful for quantitative geologic or geochemical studies. The purpose of this paper is to show: (1) that light scattering data from the L-DGO-Thorndike nephelometer can be calibrated in units of concentration and that these data are internally consistent on oceanic scales; (2) that these data contain valuable information on oceanic circulation; (3) that from these data one can derive budgets of suspended particles in different oceanic areas or for entire oceans; and (4) that together with independent data on water mass transport one can calculate fluxes of suspended particulate matter from one part of the ocean to another.

THE NEPHELOMETER AND ITS CALIBRATION

The L-DGO-Thorndike photographic nephelometer first used in 1964 has been improved primarily with respect to standardization of light-scattering response and its recording of water depth but it remains basically the same instrument. It consists of a white-light source, a calibrated attenuator and a camera which records water depth and both the attenuated direct light beam (E_D, direct-light film exposure) and the light scattered in a forward direction (E) through an angular range of about 8—24° by particles in the water. The data are generally reported as log E/E_D or log S (for scattering). The instrument is described in detail in Thorndike (1975) and briefly in Eittreim and Ewing (1974) and in Eittreim et al. (1976). Thorndike (1975) also notes that forward light scattering is most sensitive to variations in concentration, specifically, to particulate surface area. Forward scattering is also least sensitive to variations in the nature of the scattering particles, i.e. index of refraction and particle size.

A number of other nephelometers have been used and reported in the literature including those which telemeter light scattering data to the ship in real time, e.g. the nephelometer built by the Woods Hole Oceanographic Institution group and described in the appendix to Meade et al. (1975). Despite the advantages of acquiring data in real time, these instruments are considerably more complicated electronically and often require conducting cable for their use at sea. By contrast, the relative simplicity of the L-DGO-Thorndike photographic nephelometer and the fact that it can be used on any type or size of wire, are a large part of the reason that it has been successfully lowered so many times and has gathered such a large body of data.

Eittreim et al. (1976) compare results of the L-DGO-Thorndike instrument with those of other workers which measure the volume scattering coefficient β (16°). A direct comparison of the L-DGO-Thorndike and WHOI nephelometers at a number of stations is in progress and a similar comparison with an instrument modified after one by Sternberg et al. (1974) is being undertaken.

Data from all of these optical devices suffer the limitation that they are in units of light scattering which are not directly useful to the geologist or geochemist. It is, therefore, desirable that optical data be calibrated in units of absolute concentration. This is done by taking samples of water in which the nephelometer has recorded scattering and measuring the concentration of particles independently. Examples of calibration curves for various forward scattering nephelometers are given in Beardsley et al. (1970), Baker et al. (1974), Carder et al. (1974), Sternberg et al. (1974) and Owen (1974). The first calibration of the L-DGO-Thorndike nephelometer was by one of us (S.L.E.) in which concentrations were estimated by optically counting filtered particles using a microscope (Eittreim, 1970; Eittreim and Ewing, 1972). More recently we reported a calibration by means of a gravimetric analysis of filtered water samples from a limited region of the Blake Bahama Outer Ridge and the Hatteras Abyssal Plain (Biscaye and Eittreim, 1974). This calibration (referred to here as the BBOR—HAP curve) is reproduced in Fig. 1 along with another curve similarly obtained from the Lower Continental Rise (LCR) some 800 km to the northeast of the BBOR location. Descriptions

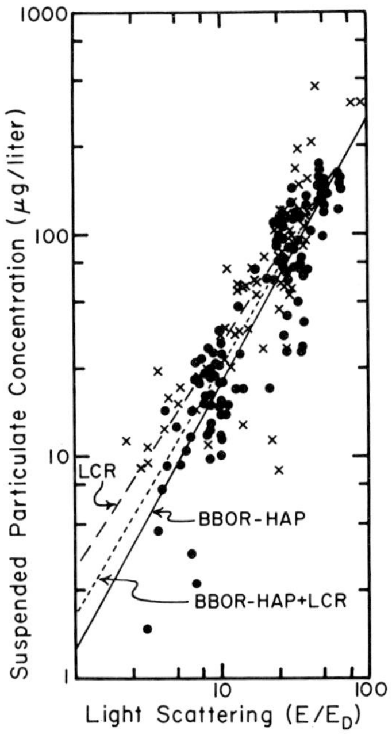

Fig.1. Calibration curves of light scattering index on the L-DGO-Thorndike nephelometer vs. suspended particulate concentrations in μg/l. Dots are data points from the Blake-Bahama Outer Ridge and Hatteras Abyssal Plain study of Biscaye and Eittreim (1974). The solid line (BBOR—HAP) is the least squares fit curve to those points. X's are data points from the Lower Continental Rise 800 km northeast of the BBOR study and the dashed line (LCR) is the least squares fit on those data. The dotted line (BBOR—HAP—LCR) is the best fit curve on all data. The BBOR—HAP (solid line) curve was used as calibration for data presented in Figs. 2—6.

of the techniques for filtration and gravimetric analyses are given in Biscaye and Eittreim (1974) and in Brewer et al. (1976). Data on the nephelometer profiles of the LCR study and their relation to hydrographic characteristics are given in Eittreim et al. (1975).

The equation in which Y = concentration (μg/l) and X = scattering (E/E_D) and r = correlation coefficient of the BBOR—HAP curve is:

$$\log Y = 1.9 \log X + 0.13 \qquad (r = 0.91) \qquad (1)$$

that of the LCR curve is:

$$\log Y = 1.0 \log X + 0.50 \qquad (r = 0.84) \qquad (2)$$

and of both sets of data taken together is:

$$\log Y = 1.12 \log X + 0.28 \qquad (r = 0.88) \qquad (3)$$

The expressions are given and figures plotted on a log—log basis for graphic convenience in compressing the ranges encountered in both parameters. Linear equations and plots yield comparable correlation coefficients. An indication of the differences between these calibrations expressed in eqs. 1—3 is given by the range of concentrations which correspond to values of light scattering from the lower and upper ends of the spectrum which are encountered in the Atlantic. For a scattering index (log E/E_D) of 0.1, such as is encountered in the clearest regions of Atlantic midwater, the BBOR—HAP curve yields 2 μg/l, the LCR curve 4 μg/l and the curve from the combined data yields 2.5 μg/l. No data in Fig.1 occur in this range since waters with these characteristics were not analyzed by both techniques in the experiments on which Fig.1 is based. Concentrations given for such low values of light scattering, of course, represent extrapolations of the calibration. For a scattering index of 1.7, such as is encountered in near-bottom water of the most intense nepheloid layers, the respective values from the three curves are 140, 160 and 150 μg/l. The greatest percentage variability between the curves thus lies toward the lower limits of light scattering.

The only other calibration curves for the L-DGO-Thorndike nephelometer from other parts of the ocean are based on preliminary data taken during the Pacific GEOSECS cruises. Calibrations based on these presently incomplete data indicate a possible increase in the range of concentrations for a given scattering index of about a factor of two or three larger than that mentioned above for the Atlantic. The correlation coefficients of these preliminary calibration curves are not as good as the two we have in the Atlantic. These calibrations indicate, however, that the range of concentrations derived from a given scattering index that will be encountered in different parts of the world ocean may be on the order of a factor of four or five. The range of actual measured concentrations encountered in the Atlantic is greater than an order of magnitude (Biscaye and Eittreim, 1974; Brewer et al., 1976) and, for vertically integrated concentrations (standing crops), approaches two orders of magnitude. We therefore conclude that it is a worthwhile exercise to apply our best present estimates of what these optical data mean in terms of actual concentrations to the body of nephelometer data in the Atlantic. We have chosen to use the BBOR—HAP curve of Fig.1 (eq. 1) rather than that combined with the LCR curve, because: first, it is the curve on which at present we have the best correlation coefficient; second, adding the LCR data does not significantly improve the geographic coverage represented (the areas are only 800 km apart and are within similar sedimentary and hydrographic regimes); and third, the differences between the curves is trivial compared to all sources of error.

In addition to the geographic limitations of the calibration used, another factor is that the calibration has been made for, and here applied to, the lower portion of the water column. The practical reason for this is that water samples taken so far in our calibration program have been those on which excess radon, a near-bottom phenomenon, has also been measured. Given the fact that the proportion of biogenic to non-biogenic particles is higher in the surface water and upper water column, we recognize the possibility that the response of the nephelometer to a given concentration of particles in the upper water column may be significantly different from its response to the same concentration of near-bottom particles. Thus, despite the minimization of these possible differences by the measurement of forward light scattering in the L-DGO-Thorndike nephelometer, we will wait until we have separate calibration curves for the upper water column before attempting this type of study in that regime.

Support for application of our single, best calibration curve to the lower water column of the entire Atlantic comes from a comparison of the nephelometer-derived data with the concentrations in the western Atlantic measured on GEOSECS samples and reported by Brewer et al. (1976). In both that paper and in Eittreim et al. (1976) are given north—south sections through the western Atlantic showing, respectively, suspended particulate concentrations and light scattering. In the southern North Atlantic and in the South Atlantic the tracks of these sections are almost the same so the data should be comparable. Applying the BBOR—HAP calibration curve to fig. 3 of Eittreim et al. (1976), much of the intermediate (clear) water is seen to carry suspended matter concentrations of from 1 to 5 μg/l. In fig. 2 of Brewer et al. (1976), much of the same intermediate water is reported to have concentrations of $\leqslant 12$ μg/l which, in the original data, actually range from 12 down to ~ 6 μg/l. Given the uncertainties in calibration discussed above, as well as the fact that the errors in the gravimetric concentrations reported by Brewer et al. (1976) are greatest for low concentrations, we take this to be reasonable agreement between the two sets of data.

SUSPENDED PARTICULATE DISTRIBUTIONS

The nephelometer data on which distributions are given here are based on those reported in Eittreim et al. (1976). Almost all nephelometer profiles reveal decreasing values of light scattering from surface to mid-water, below which values increase, to varying degrees, toward the bottom (Fig.2). We term the level at which this minimum occurs "clear water". The suspended particulate concentration at that depth calculated from the "background scattering index" (Eittreim et al., 1976) using eq. 1 is shown in Fig.3. The depth at which clear water occurs is variable in different parts of the ocean, but a considerable degree of local continuity in the depth suggests that it is the resultant of several large-scale equilibrium processes.

The principal feature of the distribution is that the lowest concentrations of suspended particulate matter occur toward the center of both the North and South Atlantic. (We recognize that less extensive data coverage in the South Atlantic permits, rather than demands, this interpretation in the contouring of Fig.3.) Low-concentration areas underlie the zones of minimum surface biogenic productivity which occur in the central regions of the main surface circulation gyres. Higher biogenic productivity and correspondingly higher concentrations of suspended particulate matter characterize surface waters around the periphery of the ocean and across the equatorial region and this is reflected in the clear water distribution in

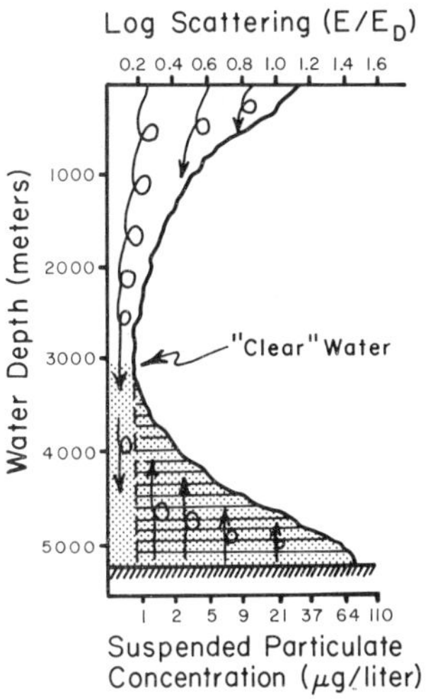

Fig.2. Typical nephelometer profile from an area with a strong nepheloid layer. The minimum in light scattering (suspended particulate concentration) is called the "clear water minimum". We define the clear water minimum as the upper limit of the nepheloid layer and all suspended matter below it as the Gross Particulate Standing Crop in units of $\mu g/cm^2$ (stippled area). The model shown here schematically assumes that all particles above the clear water minimum are falling through the water column. (The curlicue arrows represent the fact that downward settling is not presumed to be a strictly one dimensional process and that horizontal advective processes may affect the particles also.) We define the suspended matter below clear water which is in excess of the clear water concentrations as the Net Particulate Standing Crop (horizontally barred area) and assume that this represents particles mixed upward from abyssal depths (curlicue arrows upward).

Fig.3. The clear water distribution in Fig.3 is strikingly similar to the distribution of surface water biologic productivity given in milligrams of carbon per square meter per day in the Food and Agricultural Organization atlas (FAO, 1972). The addition of nutrients at the ocean margins from continental runoff, from upwelling and the seaward mixing of more productive, shallow coastal waters cause the peripheral highs in productivity. Besides being higher in biologic productivity, continental shelf waters carry higher concentrations of suspended terrigenous detritus (Manheim et al., 1970; Meade et al., 1975; Biscaye and Olsen, 1976) which also contributes to peripheral highs in the concentrations of surface water suspended matter. The fact that the suspended particulate distribution at clear water (from about 1000 to 3000 m depth) parallels the surface distribution, reflects a balance between biogenic productivity, variable rates of settling, dissolution and decomposition of the particles as they sink and are advected through the water column. The suspended matter represents a wide range of particle sizes. The effects of dissolution, decomposition, settling and horizontal advection are variable over the size range represented and a qualitative description of these processes affecting suspended particles originating in

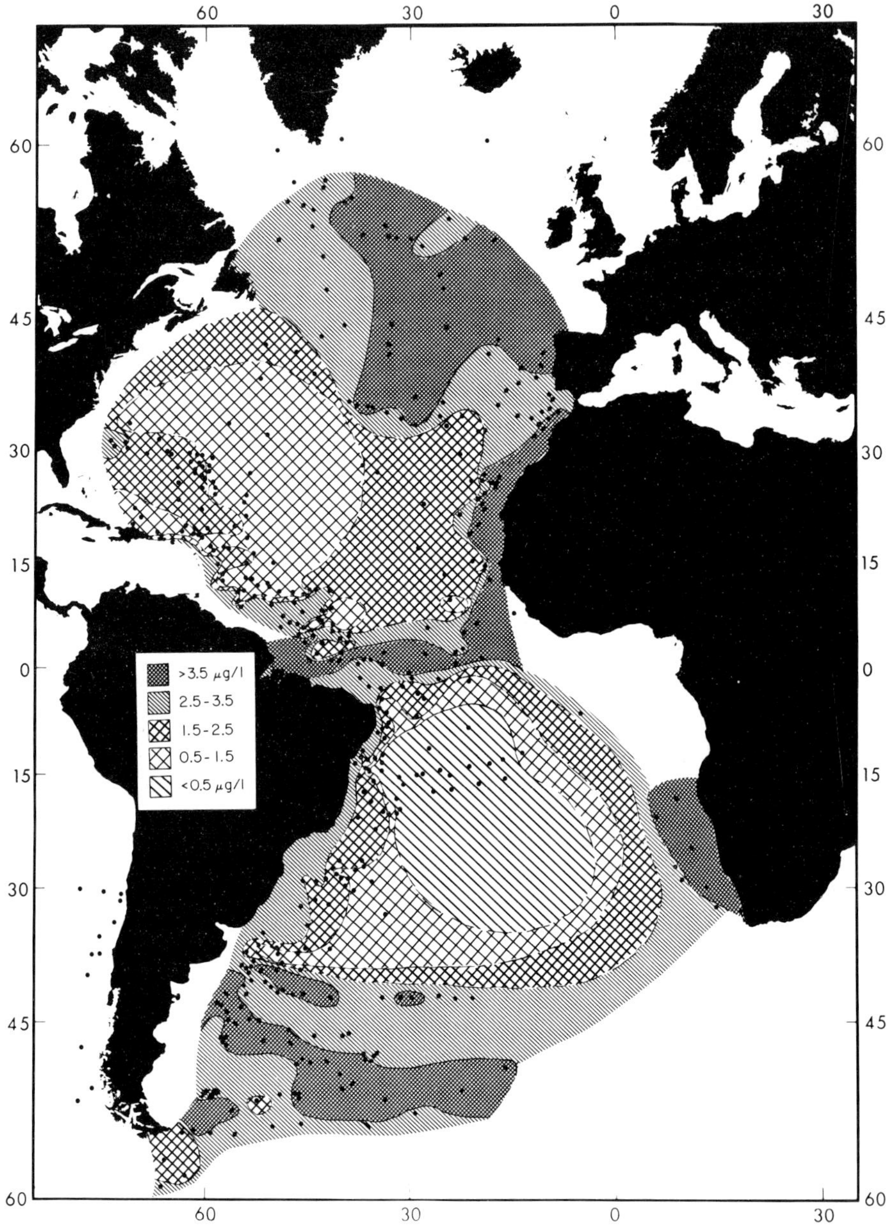

Fig.3. Distribution of the concentration of suspended particulate matter at the clear water minimum in the Atlantic Ocean.

surface water or in the upper part of the water column is suggested in Fig.2.

The arrows in Fig.2 also suggest that the clear water minimum is the resultant of upward mixing of particulates from the bottom with those processes dissolving and carrying particles downward, albeit not necessarily directly, from surface waters. These downward processes certainly continue below the clear water minimum but the particles become indistinguishable from those being mixed upward from the bottom. A vertical integration of the concentrations at each level between clear water and the bottom yields a standing crop of suspended particulate matter in units of $\mu g/cm^2$. The distribution of this integral, which we call the gross particulate standing crop, is shown in Fig.4. This map contains a number of data points in addition to those shown in Fig.3. The additional points represent old (pre-1972) nephelometer profiles which, because of instrumental limitations, are not directly intercomparable and therefore could not be used in Fig.3. These have been normalized locally at the clear water minimum using the most appropriate local background scattering index of Eittreim et al. (1976).

The distribution of gross particulate standing crops in Fig.4 is the resultant of several processes. The central portions of the North and particularly the South Atlantic have relatively low standing crops somewhat analogous to the distributions in Fig.3. Standing crops in the western basins, however, are in many places more than an order of magnitude greater than those in the eastern basins and they do not increase monotonically toward the edges of the basins but go through a maximum coincident with the axis of the western boundary current. There is thus the effect of abyssal circulation superimposed on the reflection of biogenic productivity and processes tending to increase particulate concentrations at ocean peripheries. This is seen in Fig.4 in the zones of high standing crops which correspond to the position of the western boundary currents. The cold, dense bottom water of the South Atlantic (Antarctic Bottom Water — AABW) flows into the Argentine Basin and along its western boundary northward. Suspended in the AABW and in the overlying North Atlantic deep water up to the clear water minimum are gross standing crops of more than 3,000 $\mu g/cm^2$ and which diminish progressively northward toward the equator.

The origins of the cold, dense bottom water in the North Atlantic are more complicated but a south-flowing contour current is formed of this water and is generally referred to as the Western Boundary Under Current (WBUC). In the western North American Basin it too carries suspended loads in excess of 3,000 $\mu g/cm^2$ which diminish rapidly equatorward. In contrast to the concentration of suspended particulate matter at clear water which reflects the pattern of surface water circulation, the distribution of gross particulate standing crop also reflects topographic control on abyssal circulation by the Mid-Atlantic Ridge, the Reykjanes Ridge, the Rio Grande Rise and the Walvis Ridge.

In order to see more clearly the dynamic effects of abyssal circulation on suspended particulate distributions, as opposed to those reflecting surface water circulation and productivity, we adopt the model of particle origins shown schematically in Fig.2. We recognize the limitations of this construct in that the curve described by the decrease in concentration in the upper part of the water column probably continues to decrease more-or-less

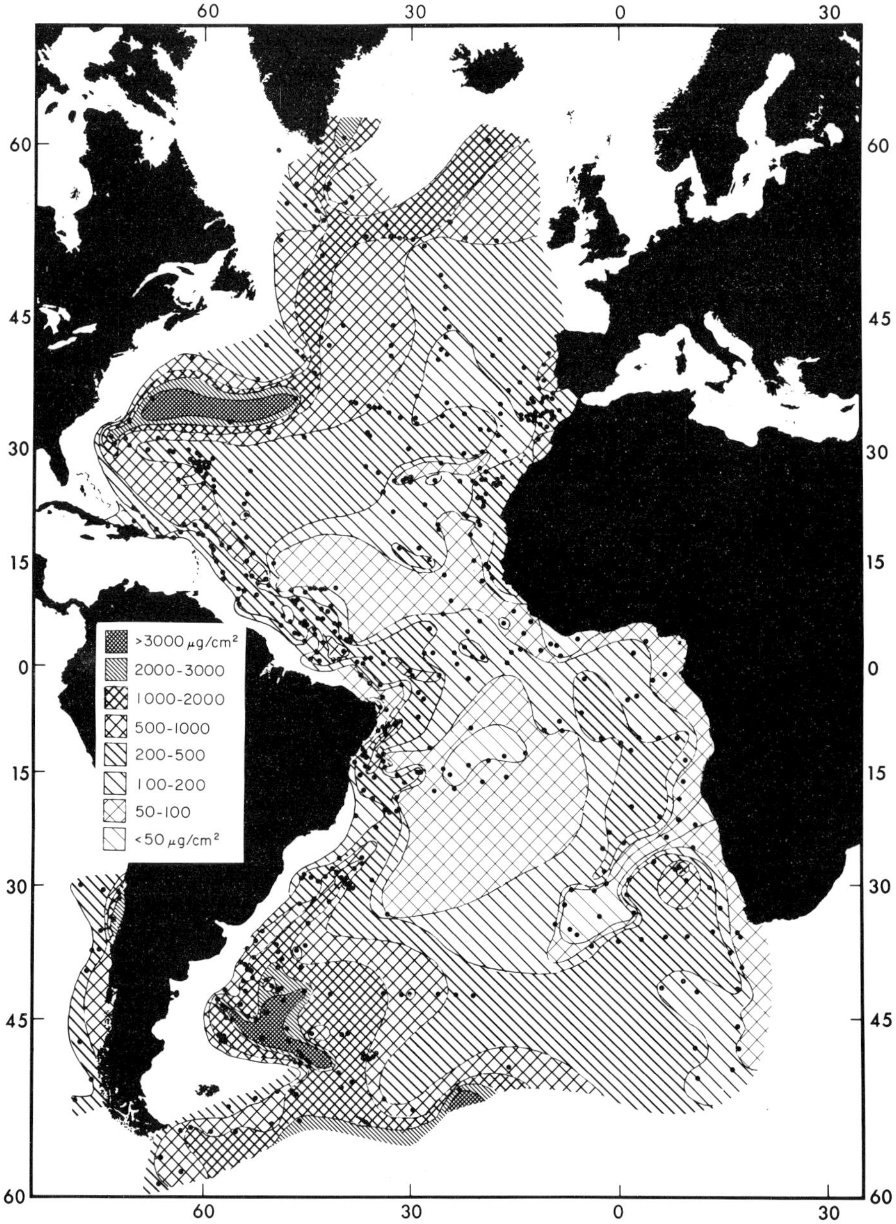

Fig.4. Distribution of the gross particulate standing crop in units of $\mu g/cm^2$ in the Atlantic Ocean.

smoothly to the bottom. The upward-decreasing curve of the nepheloid layer also probably extends somewhat above the minimum, making the observed light scattering (or concentration) minimum the resultant of the two, overlapping, asymptotic curves. Because, however, the functions describing these two intersecting curves are not known, we assume that the concentration at clear water reflects the downward resultant of biologic productivity and other surface water inputs, particle dissolution, decomposition, settling and circulation by advection and that this concentration, integrated to the ocean bottom, represents the contribution to the gross particulate standing crop of these processes. If this contribution (stippled area without cross hatching below clear water in Fig.2) is subtracted from the gross, we obtain the net particulate standing crop (portion of the profile both cross hatched and stippled), a parameter which, by this model, reflects the effect of processes which raise particles into near-bottom waters and maintain them in suspension. (The term "gross" and "net" particulate standing crop are used here in the general budgetary sense of a total and a resultant when something is subtracted from it.) These processes may include the injection of particulates into the lower water column along continental slopes and rises by turbidity currents. They may also reflect direct resuspension of bottom sediments by the erosive action of bottom currents. Of course, the net particulate standing crop at any point also includes particles advected from "upstream" by bottom currents. Using this model then, the net particulate standing crop reflects, by definition, primarily resuspended particles.

By the model presumed in Fig.2, the net particulate standing crop is probably minimized by overestimating the contribution by downward processes. Those processes of settling, advection, dissolution and decomposition of surface-origin particulates which are responsible for the decrease in particulate concentrations downward to clear water certainly continue to the bottom, i.e. the square stippled-only area is an overestimation of the surface-origin component. In those areas where vertical mixing processes from the bottom are greatest, the clear water minimum occurs higher in the water column, i.e. at a higher clear water concentration, again causing an overestimation of the background amount subtracted from the gross to yield the net particulate standing crop. Both of these effects, however, involve small quantities of suspended particulates especially compared with the net particulate standing crop values of the western basins.

The distribution of the net particulate standing crop is shown in Fig.5 and reflects a decrease at all points compared with Fig.4. Subtraction of the integral of clear water concentration, however, does not have much effect on the distributions in the western basins, but it reduces by a significant percentage the suspended loads in the eastern basins. Thus, the area covered by standing crops greater than, say, 3,000 or 1,000 $\mu g/cm^2$ in the western basins is only somewhat reduced, but, by comparison the total area covered by standing crops less than 50 $\mu g/cm^2$ is greatly increased. The principal feature of Fig.5 is the high burden of resuspended particulates coincident with the western boundary currents. The peripheral highs in gross particulate standing crop in the eastern Atlantic and along the equatorial belt of Fig.4 are almost completely removed in Fig.5. The particulate-rich outflow from the Mediterranean Sea, first noted by Jerlov (1953, 1961) not distinguish-

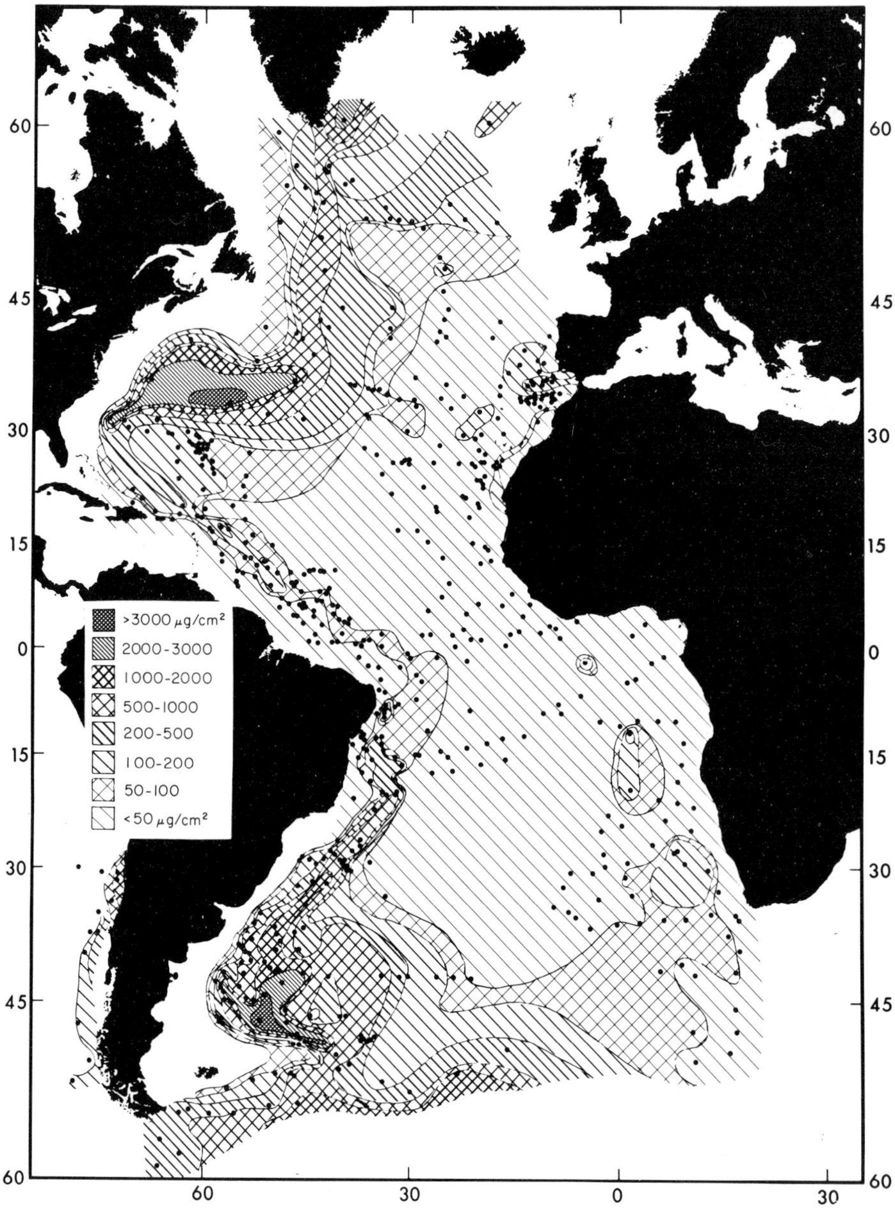

Fig.5. Distribution of the net particulate standing crop in units of $\mu g/cm^2$ in the Atlantic Ocean.

able in Fig.4, becomes apparent in Fig.5. The barrier in the eastern South Atlantic to northward spreading of the AABW presented by the Walvis Ridge is also seen in Fig.5. Thus despite the limitations of the model of particulate origins portrayed in Fig.2, the difference between the net and gross particulate standing crops emphasizes the effects of abyssal circulation in the former.

We conclude that the distribution of suspended particulates in the bottom waters of the Atlantic Ocean, the nepheloid layer, strongly reflects abyssal circulation. These large-scale dynamic effects are enhanced by a model which attempts to discern between that portion of the suspended particulate load which reflects downward transport of particles of surface water origin (largely biogenic), and that portion which results from the injection, resuspension and vertical mixing and advection of sediments at or near the sea floor.

SUSPENDED PARTICULATE BUDGETS

One can carry the use of nephelometer data beyond the distributions shown in Figs.4 and 5 by areally integrating the standing crops to yield total burdens of suspended particulates. To demonstrate the order of magnitude of these suspended loads, we have done this for a limited portion of the data. We have also calculated horizontal fluxes of suspended particles in the nepheloid layer by combining our data with published data on advective transport. Given the present limitations on our nephelometer calibration (geographic applicability as well as restriction to the lower water column) and uncertainties in the settling rates of particles, as well as uncertainties in calculated geostrophic transports, we feel that our calculated budgets and horizontal fluxes are sufficient to demonstrate orders of magnitude. To attempt details beyond this, such as calculating downward fluxes, is not warranted at this stage.

The net particulate standing crops of Fig.5 have been measured areally for most of the basins on equal area projections. The net, rather than gross, particulate standing crops were used in order to isolate that portion of the suspended particulate load resulting from near-bottom processes of resuspension and injection and therefore associated with abyssal horizontal water motions. Boundaries of the "basins" were chosen as much as possible with regard to topography. Some of the boundaries, however, are considerably less obvious than are the Mid-Atlantic Ridge and the Rio Grande Rise. The boundaries used for the several basins and areally integrated net standing crops (called net particulate loads) are shown in Fig.6. The values vary by more than an order of magnitude from $2 \cdot 10^6$ metric tons for the Guyana Basin (the long, narrow basin which includes the Demarra Abyssal Plain northeast of South America) to $52 \cdot 10^6$ tons for the North American Basin. Net particulate loads are almost an order of magnitude lower in the eastern Atlantic than is the case west of the Mid-Atlantic Ridge ($13 \cdot 10^6$ tons compared with $111 \cdot 10^6$ tons) the difference being ascribable to the strong boundary currents along the western margins.

As with the net particulate standing crops, the net particulate loads in the western basins decrease equatorward from the Argentine and North American Basins. By the model used to distinguish between descending and resuspended

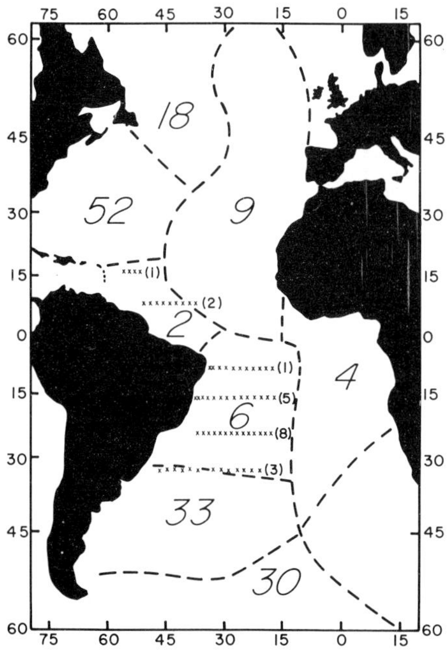

Fig.6. Total net (resuspended) particulate loads (in millions of metric tons; large numbers in italics). This represents the vertical integration of the stippled and horizontally barred area of Fig.2, integrated areally over the basins as defined by the dashed lines which generally represent natural topographic boundaries. In other words, the net particulate standing crops of Fig.5 ($\mu g/cm^2$) times the area covered by each class, summed for the whole basin equals the net (resuspended) particulate loads. Lines of X's at 32°S, 24°S, 16°S, 8°S, 8°N and 16°N are station locations for the sections across which abyssal fluxes of suspended particulates have been calculated from net particulate standing crops (Fig.5) and volume transports of AABW (from Wright, 1970). Net northward particulate fluxes are given in parentheses at the eastern end of each section in millions of metric tons per year.

sediment (Fig.2) these net particulate loads represent the total quantities of resuspended (or bottom-injected) particles. The equatorward decrease, therefore, represents either a diminishing capability of bottom currents to resuspended sediment, or the progressive fallout of particles injected most abundantly into the nepheloid layer in the North American and Argentine Basins. That distinction, however, cannot be made with the data presently available.

By either hypothesis however, bottom currents are transporting suspended matter equatorward. An estimate can be made of the rates at which this abyssal transport is occurring by using the standing crop data in conjunction with estimates of advective transport to yield horizontal particulate fluxes. We have used the values of AABW transport calculated on IGY data (Fuglister, 1960) by Wright (1970). Calculations of geostrophic transport are strongly influenced by choice of the level of no motion. By limiting ourselves to a set of north—south transports calculated by a single worker we at least availed ourselves of an internally consistent set of data, albeit at variance with transport values which might be calculated by someone else who might choose a different level of no motion. The sections are indicated by east—west rows of points (marking the IGY hydrographic stations) on Fig.6. Our method was as follows.

Wright's station positions (1970) were plotted on our Fig.5 and an average net particulate standing crop estimated for the interval between each station. Wright (1970) took the 2° isotherm as his level of no motion, i.e. the upper limit of AABW, and we subtracted the depth of this discontinuity from the average water depth between hydrographic stations to obtain average AABW thickness. This thickness is not precisely equal to the height above bottom of clear water over which the net particulate standing crop was calculated. The difference in quantity of suspended material however between clear water and the 2° isotherm is not significant. The average net particulate standing crop between stations was divided by the average AABW thickness to obtain an average net concentration of resuspended particulate matter in the AABW between each pair of hydrographic stations. In his fig.3, Wright (1970) gives a histogram of the northward and southward transports of AABW between each pair of stations as well as the net northward transport across each east—west section. We multiplied the transport increment (cm^3/yr) between each pair of stations by the average net resuspended particulate concentration ($\mu g/cm^3$) for that interval to yield a flux of suspended particulates (in millions of tons per year), which for each section resulted in a net northward flux, i.e. flux north minus flux south in all sections showed a net remainder to the north. These data are listed in Table I along with the net northward transports given by Wright (1970). The net northward suspended particulate fluxes across each section are also shown on Fig.6.

TABLE I

Northward volume transports and net (resuspended) particulate fluxes in the western Atlantic

Section	Volume transport* (10^6 m^3/sec)	Net northward resuspended particulate flux (tons/yr)
16°N	1.4	0.6
8°N	2.7	2.2
8°S	2.8	1.3
16°S	2.3	5.0
24°S	6.4	8.1
32°S	5.1	2.5

* Net northward transports of AABW from Wright (1970). See Fig.6 for the location of stations and sections.

Except for the flux at 32°S, the values generally decrease northward. Several reasons for this exception can be imagined, including the fact that Wright shows less northward flow at 32°S than at 24°S. In that the Vema Channel constitutes the primary, if not the only passage through the Rio Grande Rise for AABW northward from the Argentine to the Brazil Basin at present, the increase in net northward flow at 24°S (north of Vema Channel) compared to that at 32°S, if real, is puzzling. Perhaps the 32°S section missed some northward transport through the Hunter Channel at the eastern end of the Rio Grande Rise, a possibility suggested by Burckle and Biscaye (1970). (Wright's fig.2 indicates that he only used one station, A5821, at the eastern end of his 32°S section.) Whatever the explanation for the difference in water transport, we believe that the reason for the much larger discrepancy in suspended particulate transport probably lies in the averaging methods used in our calculation. That is, multiplying the volume transport times the average net concentration of resuspended particulate matter over the entire AABW column, although a reasonable first approximation, may average out the effects of differential horizontal transport over portions of the water column in which the suspended particulate matter concentration is highly variable. For example, if a large proportion of the northward water volume transport occurs in the lower water column where the suspended load is high (see Fig.2), the northward flux of suspended particles will be greater than if both particles and transport were uniform throughout the water column. Thus, although we believe that the fluxes in Table I and Fig.6 are of the right order of magnitude, they will be altered as the water column is examined in more detail. The next step in refining calculations of abyssal particulate fluxes should be calculation of vertical profiles of fluxes, along with locally derived nephelometer calibration curves.

Within the limitations of these calculated fluxes, the implication of the equatorward decrease in the net northward flux from 24°S to 8°S is that there is a deposition on the area underlying the nepheloid layer of about seven million tons per year. According to the model presented in Fig.2 this represents deposition only from material which was resuspended from the bottom downstream and is in addition to that which is being deposited by settling through the water column. (See also Note added in Proof, p. 171.)

ACKNOWLEDGEMENTS

We express particular gratitude to Maurice Ewing, Edward Thorndike and Larry Sullivan whose foresight and efforts to keep the program going over the years were responsible for there being such a large body of nephelometer data. We thank Karen Antlitz, Lawrence Carroll Jr. and Adele Hanley for technical assistance and Roger Chesselet, John Kanwisher, V.R. Kolla, Telu Li, Larry Sullivan and Edward Thorndike for helpful discussions and criticisms. This work is presently supported by the National Science Foundation under grant No. NSF-DES-74-01671 and during previous years by both the National Science Foundation and the Office of Naval Research.

NOTE ADDED IN PROOF

Edward Laine of Woods Hole Oceanographic Institution has pointed out a striking geographic correlation in the North American Basin between the maximum in net particulate standing crops ($\gtrsim$ 1000 $\mu g/cm^2$) and the position of the Gulf Stream-driven anticyclonic gyre proposed by Worthington (1975) as responsible for the major circulation in the North Atlantic Ocean. The position of the primary gyre between $\sim$ 33—41° N latitude and 41—75° W longitude may be seen in Worthington's fig.11 (circulation of water of $\theta < 4°C$) but Laine's conclusions were based on charts of the circulation of water of $\theta < 2°C$ provided personally by Worthington. The position of the gyre is almost identical in the two charts but the volume transports are much greater in the deep ($\theta < 4°C$) than in the bottom ($\theta < 2°C$) water. The intensity of the nepheloid layer and the volumes of transport in this anticyclonic gyre, compared to those in the WBUC are such that transport of both water and suspended particles in the WBUC only becomes significant south of the gyre, or approximately at the latitude of the Blake Bahama Outer Ridge. Laine's observations and conclusions are in his Ph.D. thesis (1976, Geological Effects of the Gulf Stream System in the North American Basin: MIT-WHOI Joint Program, Ph.D. thesis). A similar gyre circulation may, by analogy, also augment the resuspended sediment load carried by the AABW in the Argentine Basin.

Additional reference:

Worthington, L.V., 1975. On the North Atlantic Circulation. The Johns Hopkins University Press, Baltimore, Md.

REFERENCES

Baker, E.T., Sternberg, R.W. and McManus, D.A., 1974. Continuous light-scattering profiles and suspended matter over Nitinat deep sea fan. In: R.J. Gibbs (Editor), Suspended Solids in Water. Plenum, New York, N.Y., pp.155—172.

Beardsley, G.F., Pak, H., Carder, K.L. and Lundgren, B., 1970. Light scattering and suspended particulates in the eastern equatorial Pacific Ocean. J. Geophys. Res., 75: 2837—2845.

Biscaye, P.E. and Eittreim, S.L., 1974. Variations in benthic boundary layer phenomena; nepheloid layers in the North American Basin. In: R. Gibbs (Editor), Suspended Solids in Water. Plenum, New York, N.Y., pp.227—260.

Biscaye, P.E. and Olsen, C.R., 1976. Suspended particulate concentrations and compositions in the New York Bight. Limnol. Oceanogr., in press.

Brewer, P.G., Spencer, D.W., Biscaye, P.E., Hanley, A., Sachs, P.L., Smith, C.L., Kadar, S. and Fredericks, J., 1976. The distribution of particulate matter in the Atlantic Ocean. Earth Planet. Sci. Lett., 32: Lett., 32: 393—402.

Burckle, L.H. and Biscaye, P.E., 1971. Sediment transport by Antarctic Bottom Water through the eastern Rio Grande Rise. Geol. Soc. Am. Abstr. Progr., 3 (7): 518—519.

Carder, K.L., Betzer, P.R. and Eggimann, D.W., 1974. Physical, chemical and optical measures of suspended particulate concentrations: their intercomparison and application to the West African shelf. In: R.J. Gibbs (Editor), Suspended Solids in Water. Plenum, New York, N.Y., pp.173—193.

Eittreim, S.L., 1970. Suspended Particulate Matter in the Deep Waters of the Northwest Atlantic Ocean. Thesis, Columbia University, Palisades, N.Y., 165 pp.

Eittreim, S.L. and Ewing, M., 1972. Suspended particulate matter in the deep waters of the North American Basin. In: A.L. Gordon (Editor), Studies in Physical Oceanography. Gordon and Breach, New York, N.Y., pp. 123—167.

Eittreim, S.L. and Ewing, M., 1974. Turbidity distribution in the deep waters of the Western Atlantic trough. In: R.J. Gibbs (Editor), Suspended Solids in Water. Plenum, New York, N.Y., pp.213—225.

Eittreim, S.L., Ewing, M. and Thorndike, E.M., 1969. Suspended matter along the continental margin of the North American Basin. Deep-Sea Res., 16: 613—624.

Eittreim, S.L., Bruchhausen, P.M. and Ewing, M., 1972. Vertical distribution of turbidity in the South Indian and South Australian basins. In: D. Hayes (Editor), Antarctic Oceanology, II. The Australian—New Zealand Sector.

Eittreim, S.L., Biscaye, P.E. and Amos, A.F., 1975. Benthic nepheloid layers and the Ekman thermal pump. J. Geophys. Res., 80: 5061—5067.

Eittreim, S.L., Thorndike, E. and Sullivan, L., 1976. Turbidity distribution in the Atlantic Ocean. Deep-Sea Res., in press.

Ewing, M. and Connary, S.D., 1970. Nepheloid layer in the North Pacific. In: J. Hays (Editor), Geological Investigations of the North Pacific. Geol. Soc. Am. Mem., 126: 41—82.

Ewing, M. and Thorndike, E., 1965. Suspended matter in deep ocean water. Science, 147: 1291—1294.

Ewing, M., Eittreim, S.L., Ewing, J. and LePichon, X., 1971. Sediment transport and distribution in the Argentine Basin, 3. Nepheloid layer and processes of sedimentation. In: L.H. Ahrens, F. Press, S.K. Runcorn and H.C. Urey (Editors), Physics and Chemistry of the Earth, 8: 49—77.

FAO (Food and Agricultural Organization of the United Nations), 1972. Atlas of the Living Resources of the Sea. FAO/UNESCO, Rome.

Fuglister, F.C., 1960. Atlantic Ocean Atlas of Temperature and Salinity Profiles and Data for the International Geophysical Year of 1957—1958. Woods Hole Oceanographic Institution, Woods Hole, Mass., Atlas Series.

Jerlov, N.G., 1953. Particle distribution in the ocean. Rep. Swedish Deep-Sea Exped., 3: 71—97.

Jerlov, N.G., 1961. Optical measurements in the eastern North Atlantic. Medd. Oceanogr. Inst. Goteborg, Ser. B, 8: 1—40.

Kolla, V., Sullivan, L., Streeter, S.S. and Langseth, M., 1976. Spreading of Antarctic Bottom-Water and its effects on the floor of the Indian Ocean inferred from bottom-water potential temperature, turbidity and sea-floor photography. Mar. Geol., 21: 171—189.

LePichon, X., Eittreim, S.L. and Ludwig, W.J., 1971. Sediment transport and distribution in the Argentine Basin, 1. Antarctic Bottom Current passage through the Falkland Fracture Zone. In: L.H. Ahrens, F. Press, S.K. Runcorn and H.C. Urey (Editors), Physics and Chemistry of the Earth, 8: 1—28.

Manheim, F.T., Meade, R.H. and Bond, G.C., 1970. Suspended Matter in Surface Waters of the Atlantic Continental Margin from Cape Cod to the Florida Keys. Science, 166: 371—376.

Meade, R.H., Sachs, P.L., Manheim, F.T., Hathaway, J.C. and Spencer, D.W., 1975. Sources of suspended matter of the Middle Atlantic Bight. J. Sediment. Petrol., 45: 171—188.

Owen Jr., R.W., 1974. Optically effective area of particle ensembles in the sea. Limnol. Oceanogr., 19: 584—590.

Sternberg, R.W., Baker, E.T., McManus, D.A., Smith, S. and Morrison, D.R., 1974. An integrating nephelometer for measuring suspended sediment concentrations in the deep sea. Deep-Sea Res., 21: 887—892.

Thorndike, E.M., 1975. A deep sea photographic nephelometer. Ocean Eng., 3: 1—15.

Wright, W.R., 1970. Northward transport of Antarctic bottom water in the Western Atlantic Ocean. Deep-Sea Res., 17: 367—371.

Marine Geology, 23 (1977) 173—196

VISUAL OBSERVATIONS OF CONTEMPORARY CURRENT EROSION AND TECTONIC DEFORMATION ON THE COCOS RIDGE CREST*

BRUCE C. HEEZEN and MICHAEL RAWSON

Lamont-Doherty Geological Observatory and Department of Geological Sciences of Columbia University, Palisades, N.Y. 10964 (U.S.A.)

(Received July 1, 1976)

ABSTRACT

Heezen, B.C. and Rawson, M., 1977. Visual observations of contemporary current erosion and tectonic deformation on the Cocos Ridge crest. Mar. Geol., 23: 173—196.

Currents have eroded channels into the stratified conformable post-Miocene marls and chalks which mantle the Cocos Ridge. At topographically constricted sills, tidal motions superimposed on circulation locally results in flow velocities exceeding the minimum required for erosion and traction transport of foraminiferal ooze creating channels and moats. Direct visual observations from a submersible indicate that the erosion process is in operation at the present time.

Due to the local increase in the velocity of moving ocean water masses resulting from obstruction by seamounts, sediment has not been and at present is not being deposited on the flanks of the post-Miocene seamounts which rise from the crest and flanks of the Cocos Ridge. Such sediment joins that eroded from the moats and connecting channels and is presumably deposited along with those other components at the ends of the channels on the flanks of the Cocos Ridge.

Tectonic effects related to the subduction of the Cocos Ridge beneath Central America can be observed more than one hundred kilometers seaward of the subduction zone. Faulting and differential uplift associated with the arching of the crust seaward of the subduction zone was also visually observed from the U.S. Navy's deep submergence vehicle "Turtle" in 1975.

INTRODUCTION

Wilde (1966) recognized about a dozen small channels which he inferred had been eroded into the crest of the Cocos Ridge by locally generated tephra-laden turbidity currents. A few years later seismic reflection profiles and 3.5 kHz echograms revealed that channels of the type originally identified on 12 kHz echograms had indeed been eroded into the upper stratification of the Cocos Ridge (Truchan and Aitken, 1973). Deep sea drilling

———

* Contribution No. 2421 of the Lamont-Doherty Geological Observatory.

subsequently established a Middle Miocene age for the base of the sediments mantling the Cocos Ridge (Van Andel et al., 1973). The 200—500 m of pelagic sediments were found to consist of marl, chalk ooze and chalk. Nodular chert was encountered in the basal 10—100 m of chalk which overlays basaltic basement.

The NE—SW trending Cocos Ridge forms the northwest perimeter of the Panama Basin, the southern perimeter of said basin being formed by the E—W trending Carnegie Ridge which connects the Galapagos Islands to South America. The deepest sill leading into the Panama Basin lies at a depth of ∿2600 m between the Carnegie Ridge and Ecuador. Further to the west at 86°W, Lonsdale and Malfait (1974) employing a remote deep tow device

TABLE I

DSV "Turtle" dives on the Cocos Ridge made in 1975

Topography	Dive No.	Latitude	Longitude	Depth*	
Normal undisected ridge crest	14	04°32′N	86°43′W	1158 m	1097 m
				620 t	588 t
Deeply eroded channels	26	06°09′N	84°58′W	1829 m	1737 m
				1000 t	950 t
	12	06°06′N	84°58′W	1930 m	1792 m
				1032 t	980 t
	17	05°47′N	85°46′W	1591 m	1371 m
				850 t	735 t
Shallowly incised channels	30	07°29′N	84°00′W	1158 m	1079 m
				620 t	592 t
Seamount summits					
Tortuga Seamount	13	05°00′N	87°30′W	1036 m	351 m
				555 t	187 t
Lorraine Seamount	16	05°00′N	85°02′W	1825 m	1143 m
				981 t	613 t
Seamount flanks					
Lorraine Seamount	15	05°00′N	85°07′W	1668 m	1158 m
				896 t	620 t
Tortuga Seamount	13	—	—	—	—
Tutu Seamount	27	06°27′N	85°35′W	1801 m	1494 m
				972 t	802 t
Bielicki Seamount	29	08°43′N	84°45′W	1952 m	1614 m
				1049 t	867 t
Seamount moat	28	06°50′N	85°20′W	1951 m	1860 m
				1064 t	999 t
Faulted ridge crest	36	06°58′N	84°13′W	1452 m	1340 m
				800 t	720 t
	37	07°37′N	84°09′W	1930 m	1565 m
				1032 t	814 t
Uplifted eroded fault block	31	07°33′N	84°02′W	657 m	644 m
				351 t	344 t

*Depths in meters (m) and tau's (1 tau = 1/400 sec).

observed channel erosion and deep sea barchan dune development at the mouth of an erosional channel which leads north from the 2300 m central saddle area of the Carnegie Ridge.

Early in 1975 we had the opportunity of employing the U.S. Navy's deep submergence vehicle "Turtle" in an investigation of sedimentary, tectonic and volcanic features of the Cocos Ridge (Table I; Fig.1).

Much of the ridge crest lies within the present 2000 m depth capability of "Turtle". A restudy of 3.5 kHz echograms previously obtained on the Cocos Ridge focused on those features which could be reached with a 2000 m submersible, which gave prospect of significant visual effects and which were sufficiently concentrated to ensure identifiable contrasts on $\sim$2 km transects.

NORMAL CREST OF THE COCOS RIDGE

Excluding the fairly local occurrences of channels, seamounts, moats and elevated blocks, the remainder of the Cocos Ridge maintains a remarkably uniform pattern of conformable internal acoustic stratification and a smooth gentle surface topography. Although the nature of the internal reflectors seen

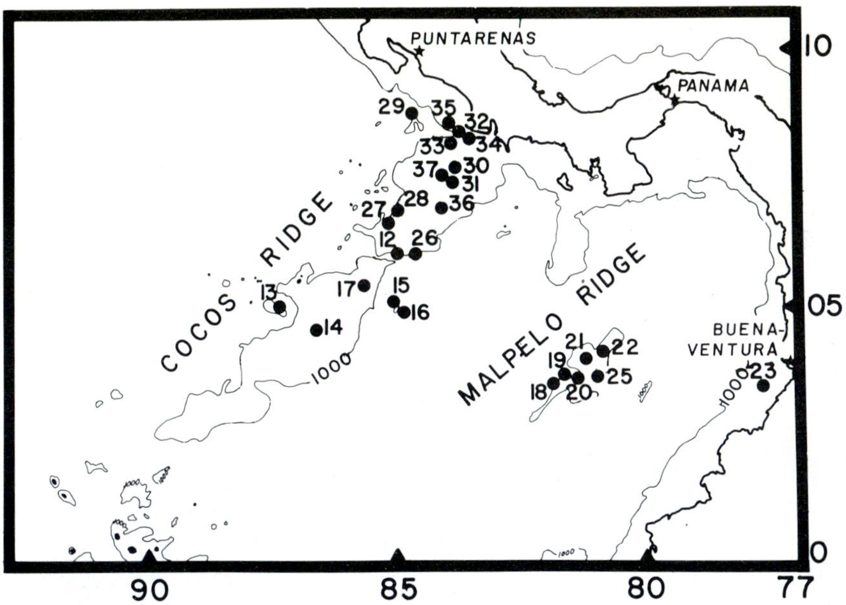

Fig.1. DSV "Turtle" made twenty-five dives in the Panama Basin during March and April 1975. The fifteen dives reported in this paper were devoted to bottom processes occurring on the crest of the Cocos Ridge and associated seamounts. On six dives similar problems on the Malpelo Ridge were studied. One dive explored the floor and walls of a large submarine canyon off Colombia and on four dives contemporary subduction activity was observed on the floor of the Middle America Trench. Depths in tau's (1 tau = 1/400 sec).

in the seismic recordings is problematical, widespread ash falls have been shown to be responsible for some of the shallow reflectors recorded on 12 kHz echograms (Worzel, 1959; Ewing et al., 1959). Neither the seismic reflection profiles, the higher resolution shallow penetration 3.5 kHz echograms nor the still higher resolution 12 kHz echograms suggest the existence of any widespread erosional or depositional bedforms. The reflectivity of the smooth crest over broad areas is remarkably strong and uniform, giving few hints of otherwise undetected small features at or near the limit of resolution. Thus, it was not at all remarkable that a 1100 m dive to the normal smooth crest (Dive 14) revealed a tranquil, tracked pelagic environment without any evidence of smoothing or lineation and with a near absence of stalked filter feeders. The visual observations in fact, confirmed what could be concluded from a study of the remote recordings: the tranquil depositional environment now prevailing at the site of Dive 14 has existed unchanged since the Middle Miocene without significant periods of erosion.

CHANNELS

Although most of the Cocos Ridge crest is covered by a continuous conformable blanket of ooze and chalk (Fig.2), the 3.5 kHz echograms reveal several classes of erosional channels. One type suggests either a perennial nondepositional regime or periods of erosion in the past alternating with more recent filling. This type was considered unlikely to allow conclusive visual observations (Fig.3).

A second class of channels cuts completely through the sediments (Fig.4). Both 3.5 kHz echograms and seismic reflection profiles suggest that basement rocks outcrop on the walls of such channels (Fig.2). This class seemed to offer significant objectives for detailed in situ visual observations.

A third class of channels cut into, but not entirely through, the sediment. Internal acoustic reflectors appear to be truncated on the channel walls

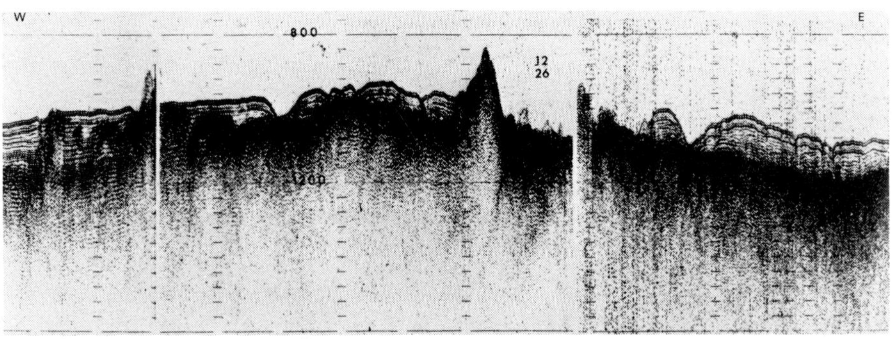

Fig.2. Seismic reflection record across the Cocos Ridge. Relatively tranquil conditions were observed on the surface of the uneroded sediment pile. Scattered outcrops of pillow lava occur in the deeply eroded areas (Dives 12 and 26). Profile extends east to west from 06°02′N 84°15′W to 06°14′N 85°35′W. Depths in tau's (1 tau = 1/400 sec).

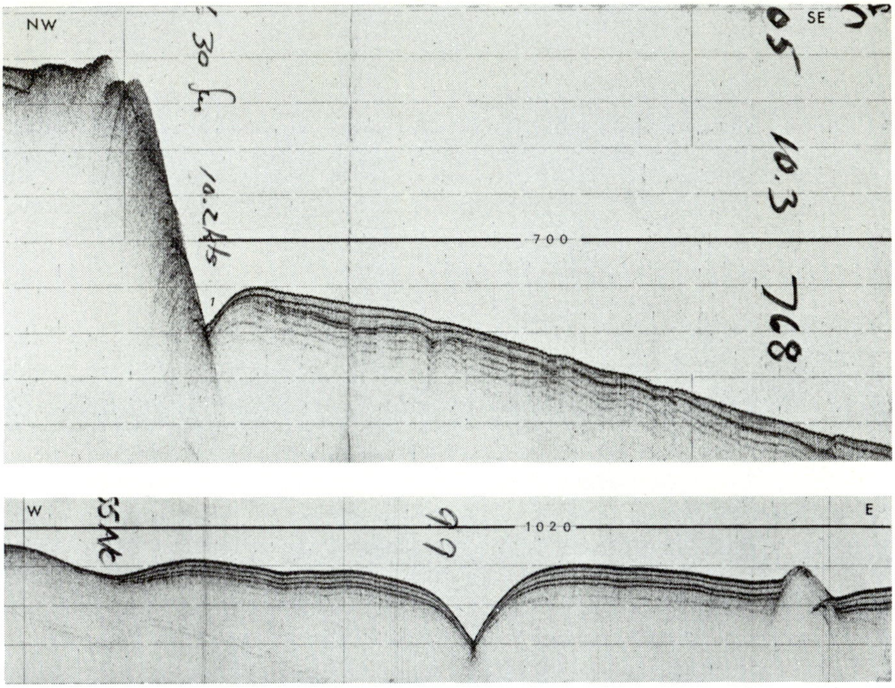

Fig.3. The upper stratification of the sediments recorded by 3.5 kHz echograms reveal contemporary processes of erosion and restricted deposition. Upper profile is part of the Coiba Ridge in the Panama Basin. Horizontal grid lines at intervals of 20 t. Profile SE to NW from 06°35′N 81°37′W to 06°39′N 81°42′W. Lower profile is an area of the Cocos Ridge which includes the site of DSDP hole 158 (left middle). The channel in the center shows restricted deposition in the latest geological time. We did not dive on such features in the belief that bottom features would be too subtle to yield significant visual results. Profile east to west from 07°29′N 83°44′W to 07°29′N 83°52′W. Depth in tau's (1 tau = 1/400 sec).

(Fig.5). This class of channels also appeared to offer the prospect of significant visual observations.

Isolated seamounts which rise from the ridge crest are in some cases surrounded by moats where all or most of the normal sediment blanket is absent (Fig.6). Moats are often connected to scour channels.

At the northern end of the Cocos Ridge, high angular topography and the association of modest earthquake activity indicates contemporary tectonic activity. Since plate models either assume or demonstrate that the Cocos Ridge is being subducted beneath Costa Rica, deformation and uplift of the ridge in advance of final subduction may be reasonably expected to effect channel development.

Deeply eroded channels

The first channel to be investigated, a prominent deeply eroded one ∼50 km long, had been crossed several times by R/V "Vema" (Truchan and

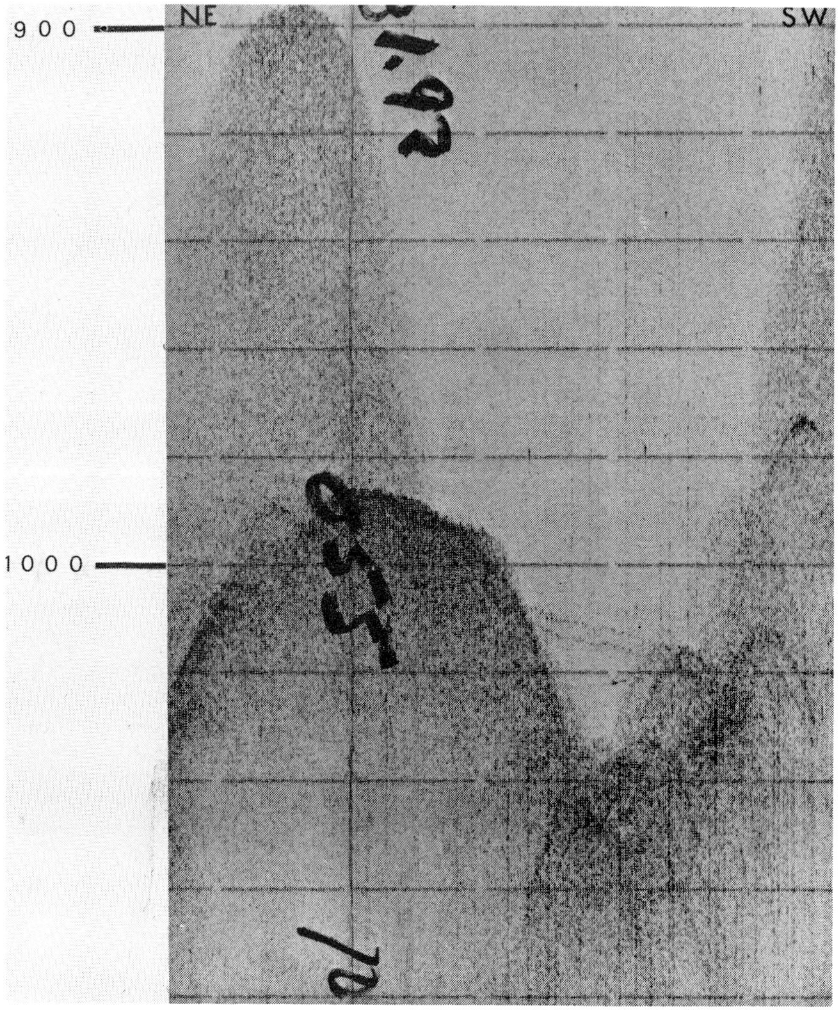

Fig.4. Echogram (3.5 kHz) across channel gives no hint of sediment cover. Scattered outcrops of pillow basalt and smoothed sediment surfaces were observed on Dives 12 and 26 on the floor of this channel. Profile SW to NE from 06°04′N 85°02′W to 06°06′N 84°59′W. Depths in tau's (1 tau = 1/400 sec).

Aitken, 1973). This channel, which crosses the Cocos Ridge from NW to SE, is joined to at least one scour moat around the base of a large seamount. The first dive (Dive 12) revealed that the channel is floored near its southeastern end by smoothed sediment and scattered outcrops of pillow basalt. Modest scour was observed around the scattered outcrops. The total lack of sediment on the outcrops confirmed the competence of contemporary currents in preventing deposition. The floor of the channel is substantially smoothed, animal "volcanoes" do not survive long in this environment and

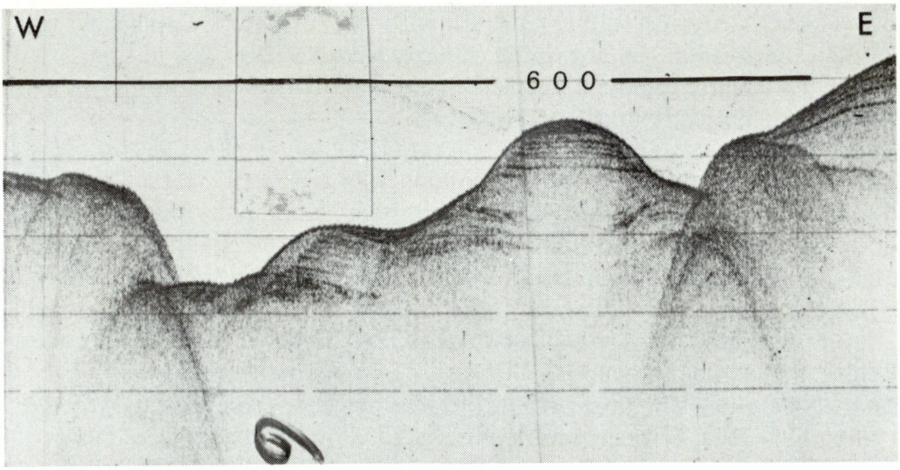

Fig.5. Scour, ripple marks and outcrops of chalk ledges were observed during Dive 30 which traversed this Cocos Ridge channel. Echogram (3.5 kHz) east to west from 07°29′N 83°58′W to 07°29′N 84°01′W.

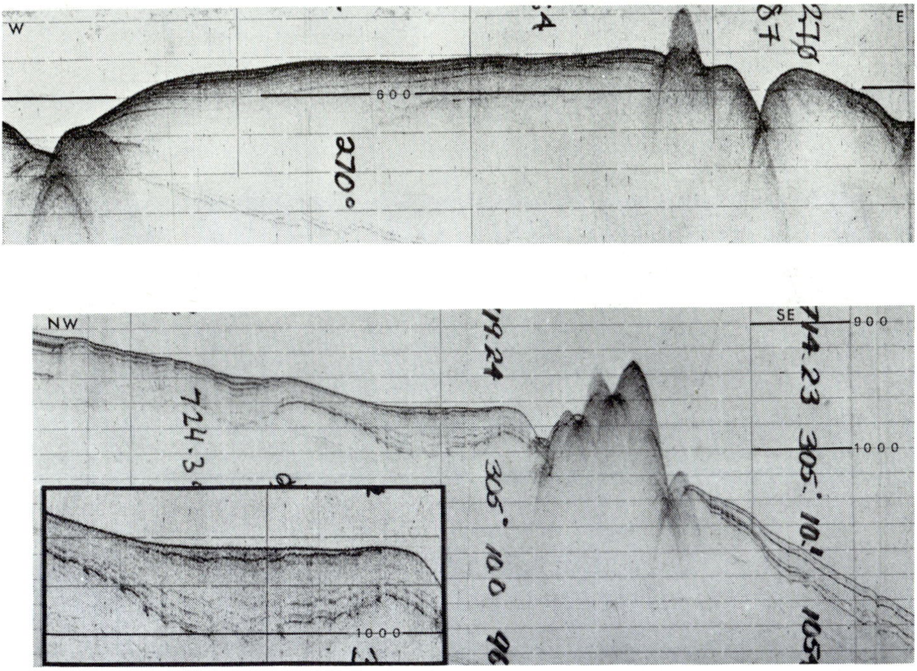

Fig.6. Erosion and redeposition on ridge crests. Upper profile: Channels are being eroded into sediments deposited on the crest of shallow northeastern part of the Cocos Ridge. The 3.5 kHz echograms indicate that chalk horizons often outcrop on the walls of the channels. Profile east to west from 07°29′N 83°52′W to 07°29′N 83°59′W. Lower profile: Sediment on the Coiba Ridge has been scoured from the vicinity of the rocky elevation creating a moat. Other sediment has ponded (insert) on a bench to the north west of the peak. Profile SE to NW from 06°26′N 81°18′W to 06° 30′WN 81° 20′W. Depth in tau's (1 tau = 1/400 sec).

those piles present tend to be elongated with the current. A ripple field was fleetingly observed at the beginning of a second dive (Dive 26) in this channel. The nearly identical bottom characteristics observed on the two dives are schematically illustrated in Fig.7.

A seismic reflection profile (Fig.2) across the channel near the site of Dives 12 and 26 indicates an almost total lack of sediment strata. The 3.5 kHz echogram (Fig.4) shows lacy hyperbolae from many small reflectors suggesting that surficial sediment is thin or absent in the channel. We had anticipated that dune forms, possibly produced by inflow over the Cocos Ridge, might occur near 2000 m in this channel. Instead it was found that the level of current activity has been below the threshold velocity necessary for the establishment of major bedforms. The sediment surface was extremely smooth. The burrow holes, tracks and feces commonly observed on tranquil bottom (Heezen and Hollister, 1971) were absent or highly subdued. The population of stalked filter feeders was greater than that seen on the normal crest at Dive 14. No dramatic bedforms were observed. Even the ripple field was small and only seen on initial approach to the bottom. Yet at the present time the current in the channel is sufficient to prevent the mantling of outcrops even by a sprinkling of ooze. Erosion, or at least nondeposition, has removed (or prevented the deposition of) several hundred meters of sediments.

Fig.7. Scour around pillow lavas on the floor of a Cocos Ridge channel. Based on two dives (Dives 12 and 26) made in March and April 1975 by DSV "Turtle". Pillow lavas appear to have little manganese cover and presumably have been denuded in the recent past by current scour through the eroded channel. Despite the fact that the seismic reflection profile (Fig.2) indicates an absence of sediment and despite the extremely fuzzy pattern (similar to that often associated with rock outcrops) detected on 3.5 kHz echograms, the two dives mostly traversed smoothed sediment bottom with only occasional scattered outcrops of pillow lavas.

The evidence that contemporary currents are eroding and smoothing the channel renders both unnecessary and unfounded Wilde's (1966) hypothesis that tephra-charged ephemeral turbidity currents eroded the channels. Instead, the process of channel erosion seems to be present-day water motions involved in oceanic circulation and, perhaps more importantly, the superimposed motions induced by the tides.

The pillow lava forms observed were not deeply encrusted by manganese which would have been the case had a stable channel floor been maintained for a long time as a non-depositional environment. This may mean that erosion has only recently reached the volcanic basement at the site of Dives 12 and 26.

Shallowly incised channels

Channels which cut into, but not entirely through, the sediment cover exposing acoustic reflectors are another class of features which appeared to give promise of significant visual observations. The 3.5 kHz echograms had suggested contemporary activity, but the possibility that the channels are relics of a more active recent past could not be completely excluded until such a channel was visually explored on Dive 30 (Figs.5 and 8).

A submersible traverse was made across the axis and up the east bank of a small channel which dissects the evenly stratified oozes of the elevated Cocos Ridge crest near 84°W (Figs.9, 10). The current observed in the channel axis flowed from N to S at about 1/3 knot. The numerous rat-tail fish consistently swam into the current. The comparatively robust benthic life of the sea floor consisted of stalked sponges bent over with the current, and occasional crabs and echinoids. Fifty meters above the channel axis the current decreased and above that depth we entered a northerly flowing current. In the transition zone between the southerly and the northerly currents, a well-defined pattern of transverse ripples was developed on the channel bank. Beneath the upper northerly current, scour and other current effects were somewhat more subdued than near the channel axis. Near the end of the dive at a depth of 1080 m, ledges of flat-lying manganese-stained chalk were observed. At that point in the dive the current ceased to flow and sediment stirred up by the submersible did not drift off but settled back to the sea floor after a few minutes. The rock ledges were devoid of sediment cover indicating that the absence of current was temporary.

CIRCULATION OVER THE SILLS

Below 2600 m the Panama Basin is totally separated from the Pacific Basin resulting in nearly isothermal conditions below sill depth. The deepest sill (2600 m) lies in the Ecuador gap between the Carnegie Ridge and the continental slope of Ecuador (Fig.11). The next higher sill is at a depth of 2300 m and lies in the central saddle area of the Carnegie Ridge near 86°W. Strong thermoclines in the Panama Basin coincide with the 2600 m and

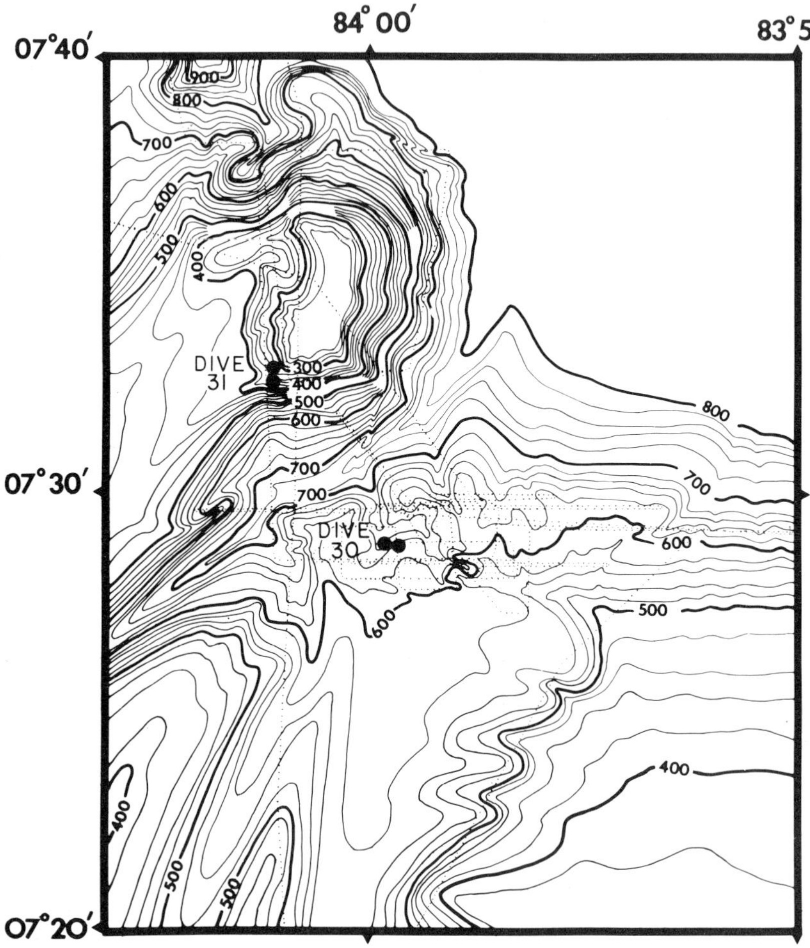

Fig.8. On Dive 31 we investigated the uplifted crest of the northern portion of Cocos Ridge. Dive 30 investigated current scour effects in a nearby scour channel. Isobaths in tau's (1 tau = 1/400 sec).

2300 m sill depths. Flow through the two gaps accounts for all the water filling the Panama Basin below 2300 m.

The Panama Basin contains $4 \cdot 10^{20}$ cm^3 of water below the 2300 m saddle depth of the Carnegie Ridge. Detrick et al. (1974), on the basis of the geothermal heat flux, estimated 100 years as the average residence time of these deep waters. Assuming a unidirectional flow below 2300 m through the combined cross-sections of the Ecuador gap and the Carnegie Ridge saddle the average velocity is 3.5 cm/sec. This value is hardly sufficient to account for erosion and traction transport, even if the estimated 6 or 7 cm/sec tidal component were to be added. Judging from the bedforms remotely observed by Lonsdale and Malfait (1974) the actual flow velocities

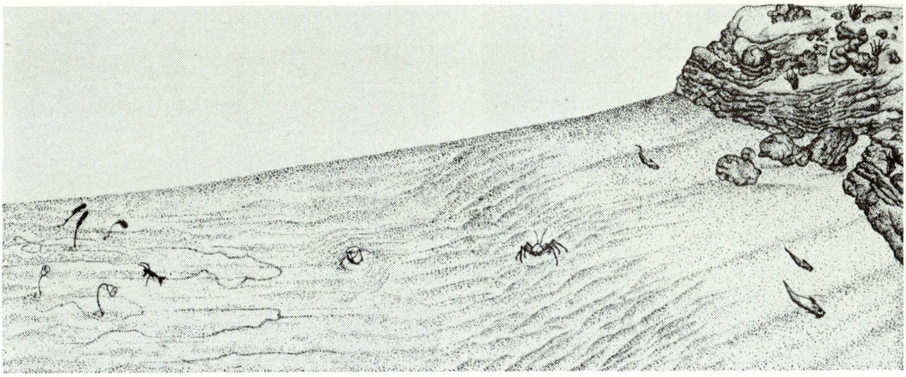

Fig.9. Sketch of principal features observed in an erosional channel at the crest of Cocos Ridge (Dive 30) (see Fig.8 for location and Fig.5 for profile of channel). Sketch is greatly compressed horizontally, the ledge on the right and channel axis to the west being approximately 2 km apart. In the floor of the channel a current flowed from north to south deflecting attached fauna and smoothing out bioturbations but leaving no scour or ripple marks. On the east flank of the channel the current diminished and changed from southerly to northerly. At this location ripple marks were prominently seen on the channel floor for a distance of a few hundred meters. Higher on the east wall current effects were more subtle than beneath the deeper southerly flowing water. The flat-lying ledges of chalk were colored by a patina of manganese but were not deeply encrusted. The ledges (at the right) were completely devoid of ooze although observers noted that the current was insufficient to carry away sediment stirred up from the adjoining ooze areas by the submersible.

across the Carnegie Ridge saddle must often exceed 30 cm/sec.

If the entire mass of Panama Basin water were to enter through a single channel similar to those found on the Cocos and Carnegie Ridges with a typical cross-section of 0.1 km^2, the resulting average velocity of 1000 cm/sec would cause rapid and deep erosion to occur regardless of the composition of the sill. Even if deep circulation depended on 10 or 20 such channels significant erosional effects would result. It is not unreasonable to imagine that in such a youthful and unstable tectonic setting inflow may have alternately occurred through a number of channels and gaps.

Water cascades from the deepest sill to the bottom of the basin but shallower sills can normally be subjected to cascading flow only to the depth of the next deeper sill. Since each successively shallower sill generally involves a smaller total mass of water and an increasingly larger sill cross-section, cascading from successively shallower and wider sills decreases until the remaining barrier ceases to cause a significant constriction. As the restrictions to flow decrease, velocities necessary for circulation drastically decrease until they become insignificant to bottom sediment and life.

Apparently the Carnegie Ridge at no time has presented a shallower sill than the Cocos Ridge. Since the Ecuador gap which now controls the flow of water into the Panama Basin is the northern end of the active Peru—Chile

Fig.10. Current scour on the Cocos and Malpelo Ridges. The rocks seen in the picture at the upper right are lithified Miocene cherty oozes. The ripples seen in the upper left are in isolated patches between rock exposures near the crest of the elevated block (Dive 31). The two photographs (Dive 30) in the middle tier are from a scour channel which lies near the base of the seamount illustrated above. Robust stalked sponges are abundant, with assorted small scour moats. The ripple pattern seen in the photograph at the center right occurs between the southerly flowing deep water and northerly flowing upper water

Trench subduction zone there is a real question as to the stability of this sill with time. The even sediment stratification over most of the Cocos Ridge testifies that, with the exception of the channel area investigated in 1975, it has never served as a major sill for Panama Basin deep waters. The crest of the Galapagos spreading center near 90°W forms part of the Panama Basin perimeter presenting a sill depth of 1700 or 1800 m (Fig.10). Since the spreading center is a dynamic growth feature of the earth's crust this sill may fluctuate significantly with time.

The dune field on Carnegie Ridge subjected to the deep-tow survey by Lonsdale and Malfait (1974) lies near the depth at which the flow cascading over the Carnegie Ridge saddle must leave the bottom and flow over the denser water which enters the Panama Basin through the Ecuador gap. The barchan dunes must terminate immediately to the north of the survey area, just as the dune fields below the Mediterranean outflow current end when that current leaves the bottom west of Gibraltar to become an interflow in the eastern Atlantic (Heezen and Johnson 1969). The seismic records reproduced by Lonsdale and Malfait (1974) show complete denudation only in a small area of the saddle crest. Below the axis of the channel leading from the saddle to the dunes most of the Pliocene and Miocene chalk sequence still remains. For the above reason Lonsdale and Malfait (1974) infer that the erosion of the saddle only began in the Middle Pliocene. Their speculation that uplift of the Carnegie Ridge was the cause of the commencement of erosion seems unlikely. It would appear that if the Carnegie Ridge had been considerably deeper it would have been subject to a greater volume of cascading flow particularly if its saddle were deeper than the Ecuador gap.

If cascading flow occurs through the Cocos channels, dunes could be expected to develop between the 1900 m depth of the maximum Cocos Ridge sill and the thermocline controlled by the Carnegie Ridge saddle. An examination of available echograms from the vicinity of the Cocos Ridge channels at that depth range failed to reveal any evidence of dune fields. Apparently significant cascading overflow does not occur in the Cocos Ridge channels at the present time.

THE FAUNA

The principal visual features of the sea floor are the effects of animal life, the next most frequent observations are the life itself and in most areas rocks and bedforms are the rarer observations. Except on extremely steep escarp-

---

at 1158 m depth along the east flank of the channel. The two photographs at the bottom of the page illustrate erosion of "bad lands" topography along the flanks of the Malpelo Ridge. Upper left: (31-36): 07°33'N 84°02'W, 650 m (347 t). Upper right: (31-195): 07°33'N 84°02'W, 645 m (345 t). Center left: (30-183): 07°29'N 84°00.7'W, 1158 m (620 t). Center right: (30-156): 07°29'N 84°00.7'W, 1151 m (618 t). Bottom left: (25-22): 03°40'N 81°05'W, 1874 m (1012 t). Bottom right: (25-96): 03°40'N 81°05'W, 1890 m (1021 t).

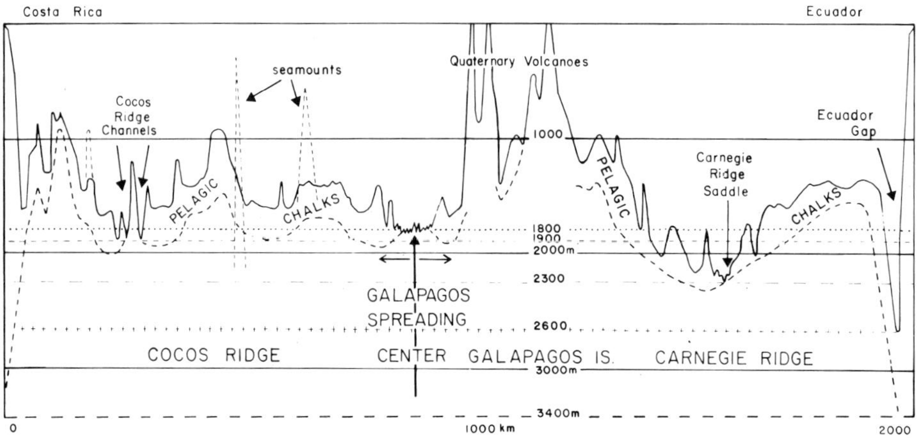

Fig.11. Profile along the crests of the Cocos and Carnegie Ridges. These ridges separate the Panama Basin from the Pacific Basin. The maximum sill depth (2600 m) is found in the Ecuador gap. The next higher sill (2300 m) lies in the Carnegie Ridge saddle. The apparent sill depth in the Cocos Ridge channel area is 1900 m. The sill depth at the Galapagos spreading center is shown as 1800 m but the area has not been surveyed sufficiently either as regards bathymetry or water structure to establish the true sill depth at this location. The general depth of the eastern Panama Basin is 3400 m.

ments and seamounts where rocks dominate, the majority of the sea floor is sediment-covered and without dramatic bedforms; and it is the bottom life and its activities which reveal the principal visual information on the physical environment.

The Panama Basin is fairly rich in animal life at the 1000—2000 m depths to which we dove. We saw thousands of small bathypterid (tripod) fish. These are excellent creatures for the observer interested in subtleties of the current pattern for they invariably stand on three stiffened elongated fins and point directly into the current. Rat-tail fish swim with their mouth near the bottom and their body pointed up at an angle of 30—40°. They respond to any current by swimming directly into it but if the current stops, they immediately swim in random patterns quickly rearranging themselves into the current when it resumes. Both fish are large enough to be easily identified, but small enough that they are often difficult to photograph, particularly the tripod fish which are of soft coloration. In the channels where current activity was rather obvious from other physical indications, there is also a greater abundance of crabs. If the tripod fish and rat-tail fish are the best and most reliable indicators of instantaneous current directions, it is the stalked filter feeders which give the best indication of continuing currents. In the channels there were abundant stalked sponges, less abundant gorgonians and occasional pennatulids. In the channel examined on Dive 30 the stalks of the sponges appeared shorter and thicker than those in other areas. Occasionally they stood erect, but more often they bent with the current sometimes even touching bottom and scribing arcs in the sediment.

The large plowers and trackers are mostly known but the makers of most of the cones, piles and holes are not certainly identified although candidates are clear from dredging results. The modification or absence of these mounds, tracks and trails can be subtle indicators of current smoothing (Heezen and Hollister, 1971).

SEAMOUNTS

Of the four conical seamounts investigated, Tutu Seamount rises from the crest of the Cocos Ridge, Lorraine rises from the south flank in the Panama Basin and Tortuga and Bielicki are located on the north flank of the Cocos Ridge in the Guatamala Basin. All are extinct volcanoes which presumably erupted after the creation of the Cocos Ridge. Pillows and lava cylinders dominated the visual scene on each of the five dive transects. Significant differences in the thickness of manganese, the presence or absence of ripple fields, variations in the amount of sediment accumulation on benches and differences in the degree of fracturing were the principal variations observed in an otherwise continuous expanse of lava pillows and cylinders.

Photographs of ripple marks at 1000 m on an Atlantic seamount (Heezen et al., 1959), ripples at 1370 m on a Pacific guyot (Menard, 1952) and the dredging of mid-Cretaceous rudistids from Pacific guyots (Hamilton, 1956) long ago demonstrated that such isolated features seemingly of rather trivial cross-section as compared to the broad oceanic basins are subject to denudation by the motions of oceanic water masses. Certainly on empirical evidence alone, any conical seamount which rises above 2000 m in an ocean basin can be expected to be nearly devoid of sediment.

Since the ocean is, in general, a highly stratified medium, motions in the horizontal plane are normal and vertical motions are exceptional. The reduction in cross-section imposed by a seamount on the motion of ocean waters is more important than one might at first realize, for a water mass is compelled to pass around the obstruction without significant vertical excursions. If the moving waters are to continue their flow it is necessary, because of the longer path, for the waters passing around an ideal conical mountain to increase velocity by a factor of $\sim\pi/2$ or ~ 1.6.

Since a velocity of about 15 cm/sec is necessary for the erosion of foraminiferal sand, we may conclude that the maximum velocities of mid-ocean water motions must frequently reach at least 10 cm/sec. An increase by a factor of 1.6 would raise that velocity to above the minimum required for erosion.

Ripple marks in small fields of foraminiferal sand were observed near the summit of Lorraine Seamount (Dive 16). These moving sands appeared to be a permanent aspect of the area for they have extensively undercut ledges of manganese-encrusted lithified sediment. Ripples were not observed on the lower flanks of any of the four seamounts investigated. Ripple marks appear to be characteristic features of summit areas and slope break regions but do not normally occur on the lower flanks of seamounts.

The 1500 m summit of Ita Maitai Guyot in the western equatorial Pacific is capped by 200 m of post-Paleocene foraminiferal sand and nannoplankton ooze. This cap thins towards the break in slope and unconsolidated sediment is absent from the upper flanks of the guyot (Heezen et al., 1973). Drilling on the summit of Horizon Guyot Ridge (Heezen et al., 1971; Winterer et al., 1973) also penetrated a cap consisting of ~300 m of post-Lower Cretaceous foraminiferal sand. The break in slope of Horizon Guyot Ridge is also an area of current activity where cherts originally formed in the central sediment cap have been later exposed by erosion (Lonsdale et al., 1972). Mid-Cretaceous rudistid reefs at the break in slope of the summit plateaus of numerous western Pacific guyots are, after 100 m.y., still unburied by sediments (Heezen et al., 1973). The central sediment cap found on guyots and the occurrence of rippled sand on seamount summits may be the result of a common phenomenon. Sediments appear to become trapped in a local current system which carries sand back and forth across the summit but does not disperse it to the seamount flanks.

Manganese is deposited on all sediment-free submarine rock surfaces. At first it merely darkens the rocks but as growth increases, surfaces become jet black and as it thickens further shapes are increasingly altered until eventually all small irregularities are eliminated. The ability of an observer to accurately evaluate such changes not only depends upon his experience and general familiarity with bottom features but on the types of rocks involved. It can be, for instance, more difficult to estimate the extent of manganese coating on tabular rocks than it is on more irregular ones which become detectably rounded more quickly and often assume strangely subdued shapes. Manganese crusts were present on each of the four seamounts examined. On the basis of a number of geochemical measurements (Bender et al., 1970); the general deep sea rate of manganese growth has been shown to be about 1 mm/m.y. This rate apparently varies widely with water motion and proximity to sources of manganese. A mid-Miocene rock which has been continuously exposed to sea water can be expected to possess a crust at least 1—2 cm thick.

Tortuga and Tutu Seamounts (Fig.12) have the smallest thickness of manganese encrustation. Shapes are so little altered that pillow lava could be easily identified. In fact, the pillows resembled recent submarine flows we have observed on the flanks of the island of Hawaii. A greater thickness of manganese was observed on Lorraine Seamount (Dives 15—16). The undercut horizontal ledges shown in Fig.13 indicate that the manganese crusts are more resistant to abrasion by the shifting foraminiferal sand than the underlying rock on which the manganese originally began accumulating. The thickest accumulation of manganese was observed on Bielicki Seamount (Dive 29) where shapes were so completely altered that the pillow lavas were only occasionally identifiable (see Fig.16). Samples of manganese crust formed over devitrified volcanic rubble in Bielicki Seamount are about 10 cm thick and the similar subdued shapes observed throughout the dive suggest that ~10 cm is a good average thickness for the seamount. Such an

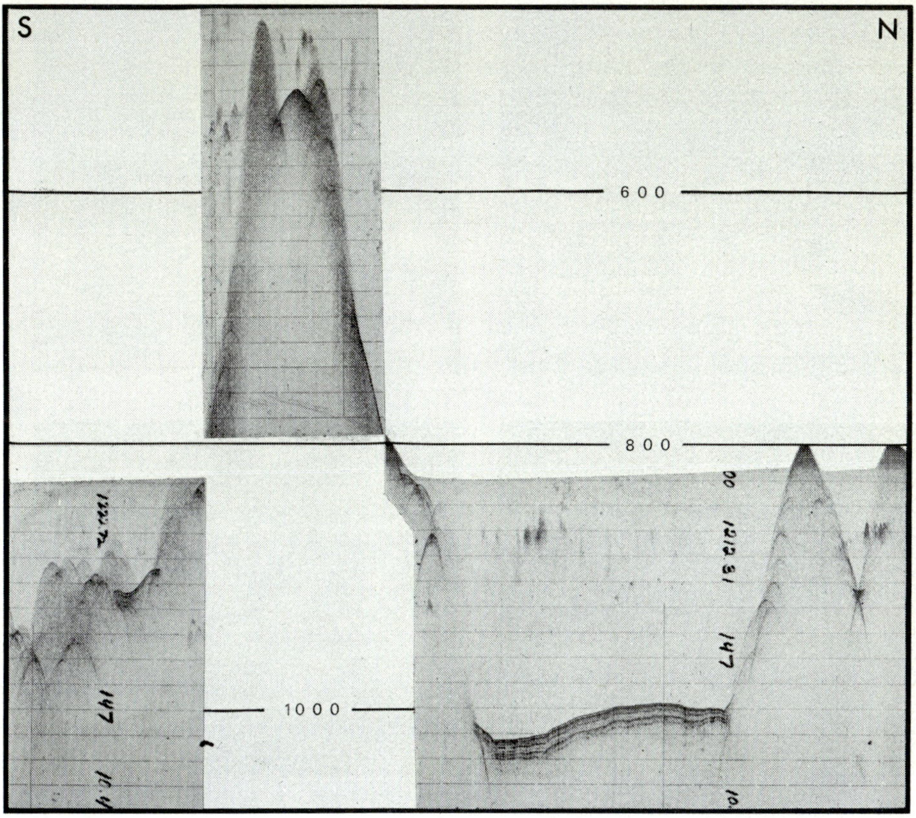

Fig.12. Tutu Seamount rising from the crest of Cocos Ridge is largely devoid of sediment cover. On Dive 27 which traversed the southern (left) flank of this seamount, steep pillow lava covered slopes alternated with steeply sloping sediment covered benches. Although differing in details the seamounts investigated at Dives 13, 15—16 and 29 have similar profiles. In each case little sediment was observed on their steep flanks. Echogram (3.5 kHz) north to south from 06°25′N 85°35′W to 06°36′N 85°38′W. Depth in tau's (1 tau = 1/400 sec).

accumulation suggests a long period during which currents have prevented the deposition of sediments. Bielicki Seamount rises from the lower flanks of the Cocos Ridge and might be somewhat older than the ridge itself.

A highly vesicular basalt was recovered at (360 m) from an outcrop near the 351 m summit of Tortuga Seamount (Dive 13). Such a vesicular texture seems unlikely to have been formed in water depths exceeding a few meters. If the subsidence curve for oceanic crust (Sclater et al., 1971) is applicable to the Cocos Ridge the summit of Tortuga Seamount would have subsided 350 m from sea level to its present depth in the past 5 m.y. Although the highly vesicular basalt from the summit of Tortuga Seamount suggests extrusion near sea level no other evidence of former subaerial environments was recognized on Dive 13.

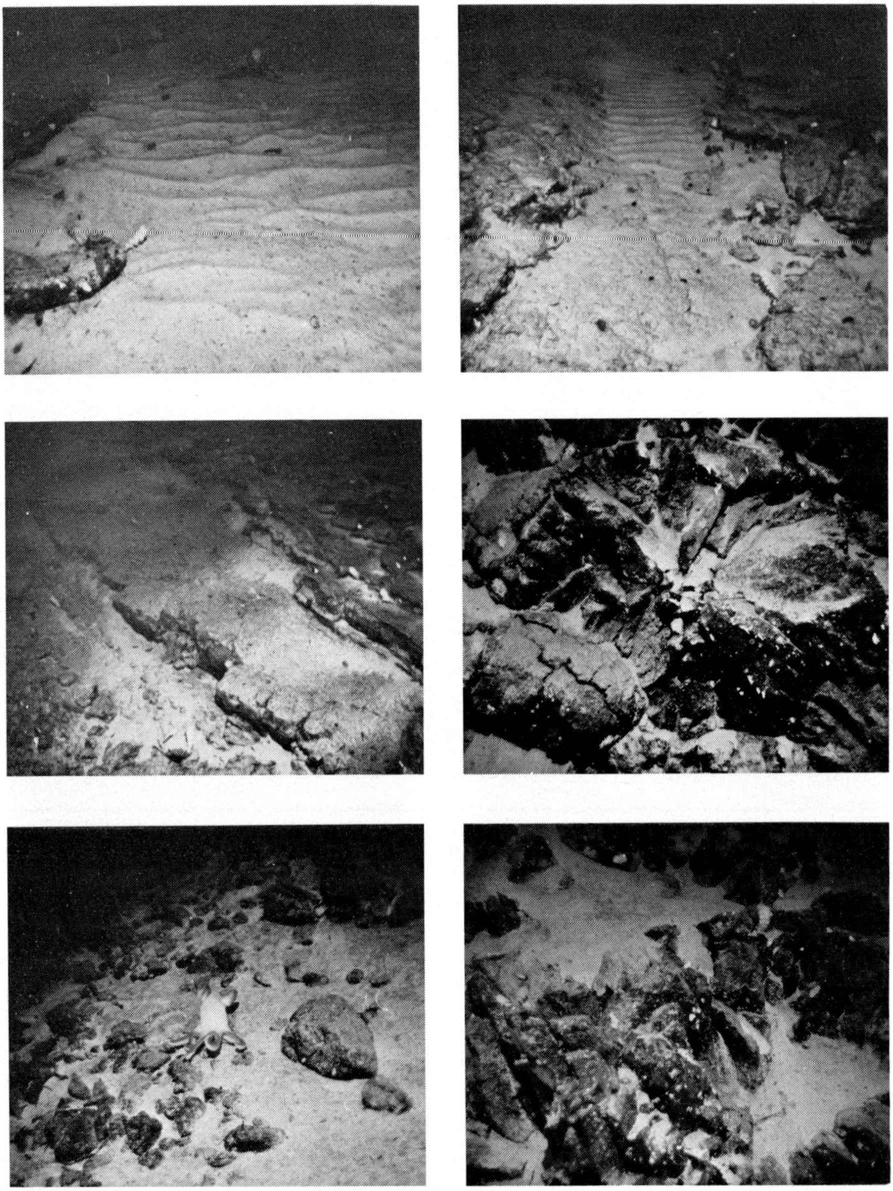

Fig.13. Characteristic views of the crest and flanks of three volcanic seamounts on Cocos Ridge. The top photographs show small patches of rippled sand which are often found at seamount summit areas. The middle photographs illustrate outcrops of bedded rock (left) and pillow lavas (right). The lower photographs show small patches of smooth sediment which accumulate on the less steep areas of the seamount flanks. Upper left: (16-175): 05°00′N 85°02′W, 1157 m (620 t). Upper right: (16-173): 05°00′N 85°02′W, 1157 m (620 t). Center left: (16-119): 04°59.8′N 85°03.2′W, 1143 m (612 t). Center right: (13-22): 05°00′N 87°30′W, 1036 m (555 t). Bottom left: (15-155): 05°00′N 85°07′W, 1371 m (735 t). Bottom right: (13-31): 05°00′N 87°30′W, 1005 m (536 t).

SCOUR MOATS

Moats surrounding seamounts and knolls are well known (Dietrich and Ulrich, 1961; Heezen and Johnson, 1963; Lowrie and Heezen, 1967). For a time during the debate concerning the origin of the moats, isostatic subsidence was proposed as a general explanation. Except for the Hawaiian moat which is accepted as a recent isostatic feature, most moats are now believed to be products of current erosion and differential deposition.

On the crest of the Cocos Ridge scour moats are occasionally associated with fairly small topographic features which rise sharply though not particularly high above the ridge crest. Thus, normal maximum velocities of water at the Cocos Ridge crest must be not less than 10 cm/sec otherwise the increase in velocity due to restricted flow would not be sufficient to cause the moating effects shown by the 3.5 kHz echograms.

Several hundred meters of pelagic ooze is lacking from the seamounts which rise from the otherwise thickly sedimented ridge crest. Since this deficit does not appear as an apron surrounding the seamount, it can be assumed that currents in the scour moat and channel system transport and disperse the missing sediments.

TILTED UPLIFTED CRUSTAL BLOCKS

A prominent topographic high on the northern Cocos Ridge was investigated on Dive 31 (Fig.8). Prior to the direct observation we had assumed that the 650 m peak was a volcano and that pillow lavas would be observed. This was not the case. The rocks exposed are not basalts but are tilted tabular consolidated chalks which dip 20° to the south. Except for an occasional patch of rippled sand the peak is swept bare. The chalks were certainly not deposited in an environment as dynamic as that which presently prevails on the peak. The history of this high broken region appears to include deposition of a sequence of chalks on the original basaltic basement in a tranquil environment followed by faulting and elevation of the present peak with consequent erosion of the unconsolidated chalks. As the peak became more of an obstacle to the movement of ocean waters, erosion increased, removing all but the most consolidated sediments. Although the peak is of different shape and composition, it is subject to erosional processes similar to those which denude the flanks of volcanic seamounts. Small patches of rippled foraminiferal sand which resemble those seen on Lorraine Seamount were observed between the tilted and beveled chalk beds at the crest of the peak.

On the broad level shallow 1100 m summit of the Cocos Ridge to the southeast of the peak, the upper stratification of the unconsolidated sediment is beveled and truncated (Fig.6) suggesting a very recent change from deposition to erosion.

ACTIVE FAULTING

A deep trough lies a few kilometers to the northwest of the uplifted block observed at Dive 31. The floor of the trough is filled with 400 m of sediment. An active fault which cuts this sediment along the south side of the trough (Fig.14) was examined on Dive 37. The sketch (Fig.15) which is greatly compressed in the vertical dimension, shows the observed characteristics of the fault. Angular talus blocks lay on the flat sediment-covered floor of the trough (1900 m) several meters beyond the foot of the talus slope. Traversing to the south for a vertical distance of 225 m we ascended this talus slope composed of cherty chalk blocks resting at their angle of repose. No sediment covered the talus and we could see far down between the blocks into darkness. Above the talus slope (1675 m) only 30 m of the vertical fault surface was exposed. Above the fault scarp the escarpment became a staircase of structural benches. Above 1585 m the sea floor levelled off. Three hundred meters beyond the edge of the upper bench, current scour diminished and the sea floor was covered by recent ooze. Discontinuous and sometimes cantilevered manganese crusts up to about 1 cm thick occur on the flat-lying chalk beds above the lip of the cliff. Nondeposition has been of considerable duration in this location. This fault has apparently been active for a considerable period of time.

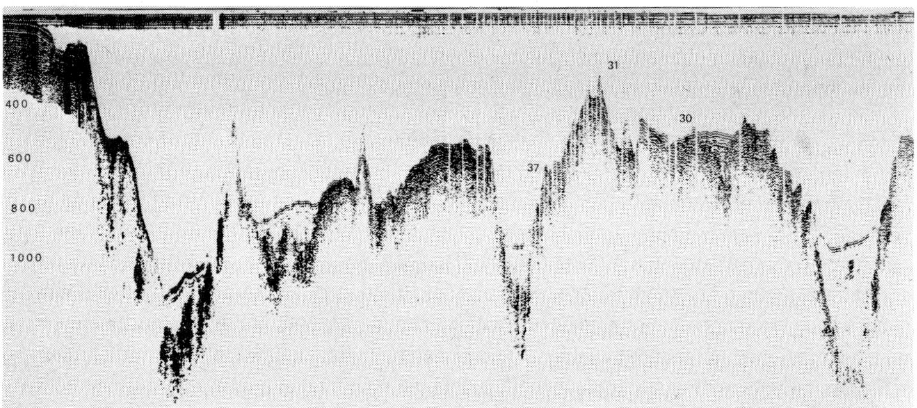

Fig.14. This seismic reflection profile from Costa Rica south across the Middle America Trench and across the adjacent northern Cocos Ridge reveals significant tectonic and erosional features. The erosional channel on the elevated crest of Cocos Ridge was investigated during Dive 30. The elevation investigated on Dive 31 exposes tilted chalk uplifted to its present erosional environment following deposition in a deeper, more tranquil environment. A fault cuts through the sediment in the depression to the left of Dive 37, where the sediment-free talus slope of an active fault was observed (Fig.15). Dives in the subduction zone of the Middle America Trench are described elsewhere. Profile South to North from 7°N 84°W to 9°N 84°W. From Expedition Iguana 3 of the Scripps Institution of Oceanography. Depths in tau's.

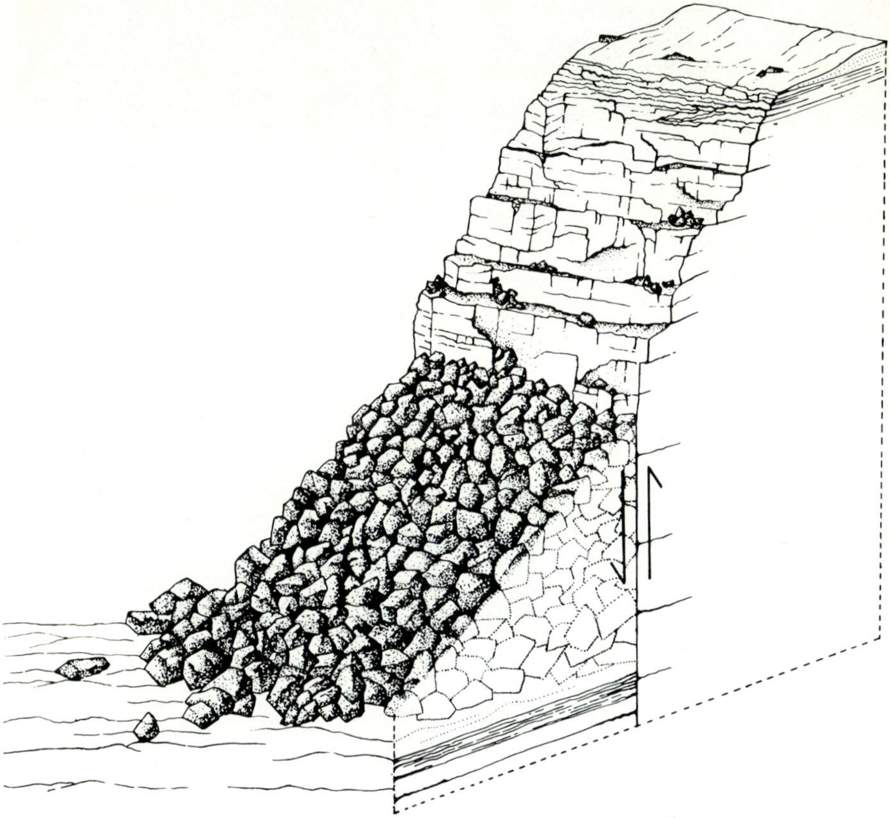

Fig.15. Sketch of fault scarp and talus on wall of tectonic depression at northern end of Cocos Ridge based on observations made during DSV "Turtle" Dive 37 in April 1975. The talus and the fault scarp are composed of Miocene cherty chalk. The fault is still active.

The faulting and uplift appear to be related to tectonic activity associated with the impending subduction of the northern part of the Cocos Ridge. As oceanic crust is carried towards a subduction zone it is broadly arched and somewhat elevated. As it comes closer to the outer wall of the trench minor vertical dislocations associated with the arching grow larger and more faults appear. On descent to the trench floor these minor dislocations grow into major faults which cut deeply into the oceanic crust.

Bielicki Seamount rises from the outer wall of the Middle America Trench near or on the northern flank of the Cocos Ridge. It has presumably been within the area of minor dislocations for several million years and has just reached the zone of major faults. Bielicki Seamount is more heavily covered with manganese, a characteristic which set it apart from other volcanic seamounts of the Cocos Ridge. It also is characterized by many dislocations (Fig.16) which contrast markedly with the regular depositional lava flanks

Fig.16. Seamounts, erosional channels and benthic life. The top pair of photographs show the manganese encrusted rocks of Bielicki Seamount which lies on the descending outer wall of the Middle America Trench. Pillow lavas exposed in the floor of an erosional channel on the Cocos Ridge are seen in the photograph at lower right. Upper left: (29-19): Manganese encrusted faulted rocks on Bielicki Seamount, 08°43′N 84°45′W, 1950 m (1048 t). Upper right: (29-15): Tabular manganese encrusted rocks on Bielicki Seamount, 08°43′N 84°45′W, 1950 m (1048 t). Center left: (13-15): Vesicular pillow lavas near the crest of Tortuga Seamount, 05°00′N 87°30′W, 1036 m (555 t). Center right: (32-198): Echinoid on a smoothed portion of floor of the Middle America Trench, 08°23′N 83°51′W, 1661 m (892 t). Bottom left: (25-110): Basket star on tranquil bottom at the southern flank of Malpelo Ridge, 03°40′N 81°05′W, 1859 m (998 t). Bottom right: (12-133): Pillows exposed on floor of erosional channel on Cocos Ridge, 06°06′N 84°58′W, 1844 m (990 t).

of the other seamounts. The faulting and fracturing of Bielicki Seamount is entirely in keeping with its tectonic setting on the brink of subduction.

ACKNOWLEDGEMENTS

The authors express their thanks to the U.S. Navy's Submarine Development Group 1 for the use of the deep submergence vehicle "Turtle". Her officers and enlisted men and the master and crew of the surface vessel "Maxine D" deserve particular mention for their professional enthusiastic support. This research was supported by the U.S. Navy Office of Naval Research under Contract N-00014-76-C-0264. The sketches illustrating dive results were done by Suzanne MacDonald. Daniel J. Fornari, Dale Chayes, Nelson Letourneau and Michael Condit, observers on one or more dives, assisted the authors in evaluating dive observations. Leslie Rosenfeld and Ray Lynde provided assistance in the preparation of the paper. Marek Truchan and Thomas Aitken provided valuable suggestions. Alexander Malahoff proposed the in situ investigations on the Cocos Ridge and provided continuous encouragement during the course of the investigations.

REFERENCES

Bender, M.L., Ku, T.L. and Broecker, W.S., 1970. Accumulation rates of manganese in pelagic sediments and nodules. Earth Planet. Sci. Lett., 8: 143—148.

Detrick, R.S., Williams, D.L., Mudie, J.D. and Sclater, J.G., 1974. The Galapagos spreading center: bottom-water temperatures and the significance of geothermal heating. Geophys. J. R. Astron. Soc., 38: 627—637.

Dietrich, G. and Ulrich, J., 1961. Zur Topographie der Anton-Dohrn-Kuppe. Kiel. Meeresforsch., 17(1): 3—7.

Ewing, M., Heezen, B.C. and Ericson, D.B., 1959. Significance of the Worzel deep sea ash. Proc. Natl. Acad. Sci. U.S.A., 45: 355—361.

Hamilton, E.L., 1956. Sunken islands of the mid-Pacific mountains. Geol. Soc. Am. Mem., 64: 97 pp.

Heezen, B.C. and Johnson, G.L., 1963. A moated knoll in the Canary Passage. Dtsch. Hydrogr. Z., 16: 269—272.

Heezen, B.C. and Johnson, G.L., 1969. Mediterranean Undercurrent and microphysiography west of Gibraltar. Bull. Inst. Oceanogr. Monaco, 67: 95 pp.

Heezen, B.C. and Hollister, C., 1971. The Face of the Deep. Oxford Univ. Press, London, 659 pp.

Heezen, B.C., Tharp, M. and Ewing, M., 1959. The Floors of the Oceans, I. The North Atlantic. Geol. Soc. Am. Spec. Pap., 65: 122 pp.

Heezen, B.C., Fischer, A.G. et al., 1971. In: Initial Reports of the Deep Sea Drilling Project, VI. U.S. Government Printing Office, Washington, D.C., pp.17—39.

Heezen, B.C., MacGregor, I.D. et al., 1973. Oolitic Limestone on Ita Maitai Guyot, Equatorial Pacific: DSDP Site 202. In: Initial Reports of the Deep Sea Drilling Project, XX. U.S. Government Printing Office, Washington, D.C., pp.97—102.

Laird, N.P., 1971. Panama Basin deep water — properties and circulation. J. Mar. Res., 29: 226—234.

Lonsdale, P. and Malfait, B., 1974. Abyssal dunes of foraminiferal sand on Carnegie Ridge. Geol. Soc. Am. Bull., 85: 1697—1712.

Lonsdale, P., Normark, W.R. and Newman, W.A. 1972. Sedimentation and erosion on Horizon Guyot. Geol. Soc. Am. Bull., 83: 289—316.
Lowrie Jr., A. and Heezen, B.C., 1967. Knoll and sediment drift near Hudson Canyon. Science, 157: 1552—1553.
Menard, H.W., 1952. Deep ripple marks in the sea. J. Sediment. Petrol., 22: 3—9.
Sclater, J.G., Anderson, R.N. and Bell, M.L., 1971. Elevation of ridges and evolution of the Central Eastern Pacific. J. Geophys. Res., 76: 7888—7915.
Truchan, M. and Aitken, T., 1973. Site surveys on the Coiba and Cocos Ridges in the Panama Basin. In: T. Van Andel, G.R. Heath et al. (Editors), Initial Reports of the Deep Sea Drilling Project, XVI. U.S. Government Printing Office, Washington, D.C., pp.473—495.
Van Andel, T.J., Heath, G.R., Malfait, B., Heinrichs, D.F. and Ewing, J., 1971. Tectonics of the Panama Basin. Geol. Soc. Am. Bull., 82: 1489—1508.
Van Andel, T.J., Heath, G.R. et al., 1973. Site 158. In: T. Van Andel, G.R. Heath et al. (Editors), Initial Reports of the Deep Sea Drilling Project, XVI. U.S. Government Printing Office, Washington, D.C., pp.151—230.
Wilde, P., 1966. Quantitative measurements of deep sea channels on the Cocos Ridge, east central Pacific. Deep-Sea Res., 13: 635—640.
Winterer, E.L., Ewing, J.I. et al., 1973. Site 171. In: Initial Reports of the Deep Sea Drilling Project, XVII. U.S. Government Printing Office, Washington, D.C., pp.283—334.
Worzel, J.L., 1959. Extensive deep sea sub-bottom reflections identified as white ash. Proc. Natl. Acad. Sci. U.S.A., 45: 349—355.

Marine Geology, 23 (1977) 197—215

IGNORANCE CONCERNING EPISODES OF OCEAN-WIDE STAGNATION*

WILLIAM B.F. RYAN and MARIA B. CITA

Lamont-Doherty Geological Observatory, Palisades, N.Y. 10964 (U.S.A.)
Department of Geology and Paleontology, University of Milan, Milan (Italy)

(Received August 8, 1976)

ABSTRACT

Ryan, W.B.F. and Cita, M.B., 1977. Ignorance concerning episodes of ocean-wide stagnation. Mar. Geol., 23: 197—215.

Episodes of basin-wide abyssal stagnation have occurred in the Mediterranean Sea during the "glacial" Pleistocene and on a much larger scale in the Atlantic and Indian Oceans during the Cretaceous Period. The sedimentary products of euxinification are organic-rich sapropels which have accumulated an order of magnitude more carbon during the Cretaceous than that which is present in all the known world reserves of coal and petroleum. The storage in the Cretaceous strata of excess carbon and sulfur in the form of fossilized photosynthetic substances and pyrite is thought to have led to a significant global increase in atmospheric oxygen and to an unusual sequence of calcium-rich evaporitic salts in the South Atlantic during the Aptian.

Stagnant episodes in the Mediterranean and Atlantic effectively destroyed all benthic life formerly existing on substratums underlying hydrogen-sulfide bearing anoxic bottom waters. Locally thin organic-rich strata on topographic highs near the paleo-equator of the Pacific Ocean during the Cretaceous owe their origin to an intermittently expanding oxygen-minimum zone as contrasted to total euxinification within the deep basins of the Atlantic and Indian Oceans. In the context of both land and sea areas, the late Mesozoic may have conceivably been the major "carboniferous" period of the earth since the end of the Precambrian.

INTRODUCTION

In striving to comprehend the causes of change on the surface of the earth, investigators are searching at an accelerated pace into the record of ancient environments preserved in deep-sea sediments. A significant new repository of information concerning change through time in such diverse sectors as the earth's atmosphere, its hydrosphere and the geometry and distribution of various lithospheric plates is the growing collection of subsurface cores from the Deep Sea Drilling Project. A sequence of interesting events is now being

---

* Contribution No. 2420 of the Lamont-Doherty Geological Observatory.

unravelled by the so-called paleoenvironmentalists or paleoceanographers. It extends well beyond one hundred million years and reveals some startling circumstances, which although poorly comprehended right now, might eventually be perceived as quantitatively significant perturbations in the past chemical balance of our planet.

The particular events to be briefly explored at this symposium concerning "the influence of abyssal circulation on sedimentary accumulation in time and space" are the newly appreciated episodes of ocean-wide oxygen starvation that were experienced in the latter half of the Mesozoic Era. The most significant evidence pointing towards an oxygen crisis are intervals in the drill cores of dark bituminous shale and mudstone (Lancelot et al., 1972; Berger and von Rad, 1972; Saunders et al., 1973; Girdley et al., 1974; Robinson et al., 1974; Andrews and Packham et al., 1975; Larson and Moberly et al., 1975). The composition of these so-called sapropelic deposits provides evidence that they act as unusually large receptors for elemental carbon in the form of organic substances and elemental sulfur mineralized into pyrite. The widely observed lack of mechanical churning and digestion of the sedimentary particles is a sign that during the time in which they were deposited practically all consumptive processes by bottom-dwelling animals came to a standstill. Anomalously low surface-ocean fertility is suggested by small concentrations of planktonic skeletal components and of abnormal ratios of siliceous to calcareous micro-fauna. In addition the poor state of preservation of some of the tests points to enhanced corrosiveness of ocean bottom waters. Some of these happenings and more may be related to complex biological and geochemical feedback mechanisms which came into action in response to breathing crises experienced by the ocean.

The thrust of this report is not so much to describe the nature of the organic-remains bearing sediments, but to emphasize the possible global implications of their development and the uncertainty which exists as to the processes by which they formed across such wide areas of the sea floor.

STAGNANT CYCLES

The Black Sea offers a contemporary example of a deep-sea basin lacking in dissolved oxygen everywhere from the 200-m depth range down to the center of its bathyal plain at 2200 m (Degens and Ross, 1973). The present euxinic setting commenced approximately 7000 years ago by a stratification of the water column whereby an overlying low-salinity surface layer, fed by coastal rivers and rainfall, is isolated from a denser and saltier deep layer influxing from the Mediterranean via the Bosphorous passage during the Holocene transgression (Bukry et al., 1970). The resulting stagnation is dynamically maintained by an inability of this basin to overturn its water masses and ventilate the deeper regions in oxygen vital for the consumption and decay in organic matter raining down today primarily as pelagic sediments. Winters are neither sufficiently cold nor is evaporation sufficiently high to permit waters presently in contact with the atmosphere to mix downwards and dissipate the ~2000 m thick puddle of hydrogen sulfide that has accumulated in the absence of oxygen (Deuser, 1973).

In the lack of a steady-state chemical balance whereby carbon fixed from the atmosphere by photosynthesis in the living surface layer of the Black Sea is once again returned by oxidation of sea-floor substances to carbon dioxide, the Black Sea Basin evolves into a more or less permanent geologic reservoir for carbon, leading to contents which approach 20% of the bulk sediment in the young Holocene-age sapropel (Ross et al., 1970).

Although now moderately well ventilated in all its individual deep basins, the Mediterranean was subjected in its eastern Ionian, Levantine, and Aegean Basins to perhaps up to two dozen intermittent stagnations on and off during the last 5 m.y. (Kidd et al., 1976). The more carefully documented ones have been recovered in long piston cores and belong to the "glacial" Pleistocene (0—0.9 m.y.).

In most instances the Mediterranean stagnations were brief and each lasted only a few thousand years forming dark organic layers of up to 30 cm in thickness (Olausson, 1961; Ryan, 1972; Nesteroff, 1973; McCoy, 1974). The Mediterranean sapropels are characterized by variable organic-carbon contents of barely over 1% upto 18% and average about 4%.

In terms of thickness and extent about 2 m of sapropel were deposited across an area of $0.5 \cdot 10^6$ km^2 spread over a cummulative interval of 40,000 years within the last 200,000 years. This magnitude of material calculates to a net flux of $1.6 \cdot 10^{12}$ g/year of carbon into the sedimentary reservoir — an amount which is comparable to the $1.2 \cdot 10^{12}$ g/year being tucked away continuously in pelagic sediments of all other basins* in the world's oceans.

Hence it is not difficult to imagine that if the mechanisms which induce intermittent oxygen depletion in small semi-isolated basins such as those in the Black or Mediterranean Seas were to have been activated in the distant past on a larger scale in the juvenile Atlantic, Indian or Pacific Oceans, a significant anomaly might be expected relative to the magnitude of carbon normally subtracted from atmospheric and oceanic reservoirs by other geological processes.

MESOZOIC EVENTS

Drill cores of the "Glomar Challenger" have in fact uncovered sapropelic layers as a characteristic feature of Cretaceous age strata in these larger oceans beginning with Leg 11 in 1972 and continuing with the latest drillings of Legs 43 and 44. Due to proprietary considerations this paper, however, will only discuss data which have already been published in the Initial Report Volumes (up to Leg 32) and that from a later Leg (40) on

---

* Using an area of $150 \cdot 10^6$ km^2 from the hypsometric curves of Menard and Smith (1966), an average sediment accumulation rate of 5 m/m.y. and an organic content of 0.1% (Lisitzin, 1972).

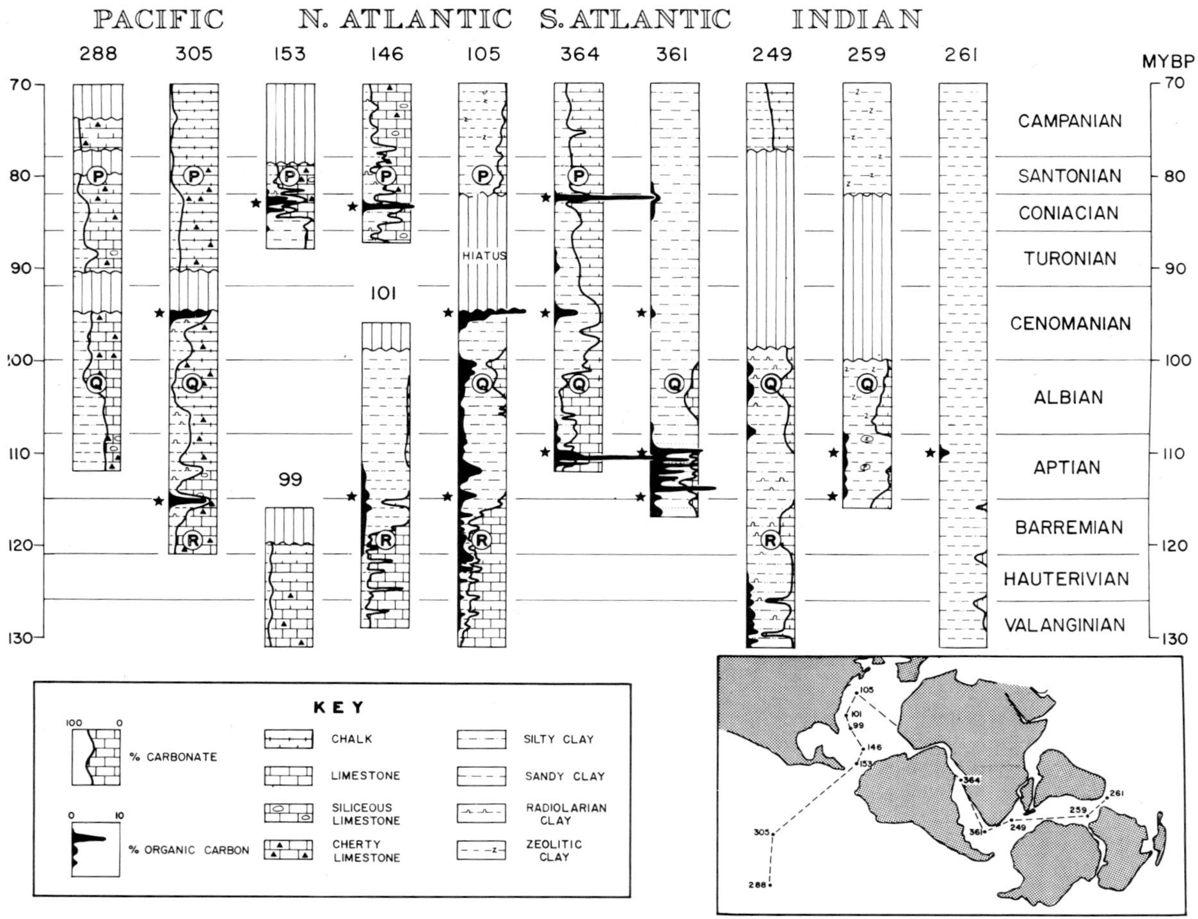

Fig.1. Lithologic correlations of stratigraphic sections of the Deep Sea Drilling Project showing time horizons accompanied by the extensive development of carbon-rich sapropels (asterisks) which are in turn bracketed by levels (*P*,*Q* and *R*) with relatively high concentrations of calcium carbonate. The map insert is a geographic reconstruction for the Aptian stage of the Cretaceous. The time scale is from Van Hinte (1976). Basinal settings outside the narrow Atlantic and Indian Oceans (Site 261) are generally devoid of sapropels whereas those on topographic highs near paleo-equatorial settings (Site 305) contain thin bituminous horizons. Organic-rich sediment is widespread within the Atlantic and Indian Oceans regardless of topographic setting except for shallow locations near the photic zone.

which one of the authors participated. A composite profile of relevant drill sites is illustrated in Fig.1 with latitudes, longitudes and depths listed in Table I.

Of interest on the lithostratigraphic cross-section are four time horizons (marked by asterisks) when extensive layers of very high organic contents were laid down — namely — the latest Barremian, the Aptian, the mid-Cenomanian, and the latest Coniacian. In some cores, such as those from the South Atlantic, the amount of carbon exceeds 20% of the bulk weight of the sediment (Foresman, 1976). In many instances, like those in the Mediterranean, the organic matter is concentrated in several discrete levels some tens of centimeters thick of pelagic origin that are interbedded in limestones, chalks and marls with a diversified benthic fauna (e.g. Sites 146, 153, 249, 305 and 364). In other cases, the organic matter is found in shales, siltstones and sandstones swept onto and across sedimentary aprons and abyssal plains by turbidity currents originating upslope on the continental margin (e.g. Sites 101, 105 and 259).

TABLE I

Drill site locations

Site No.	Topographic setting	Location	Depth (m)
99	Cat Gap North Atlantic Ocean	23°41′N 73°51′W	4914
101	Blake Bahama Basin North Atlantic Ocean	25°12′N 74°26′W	4868
105	East Coast, Continental Rise North Atlantic Ocean	34°54′N 69°10′W	5251
146	Venezuela Basin Caribbean Sea	15°07′N 69°23′W	3949
153	Aruba Gap Caribbean Sea	13°58′N 72°26′W	3923
249	Mozambique Ridge SW Indian Ocean	29°57′S 36°05′E	2088
259	Perth Abyssal Plain E Indian Ocean	29°37′S 112°42′E	4706
261	Argo Abyssal Plain E Indian Ocean	12°57′S 117°54′E	5667
288	Ontong Java Plateau W Pacific Ocean	5°58′S 161°50′E	3000
305	Shatsky Rise NW Pacific Ocean	32°00′N 157°51′E	2903
361	Cape Basin South Atlantic Ocean	35°04′S 15°27′E	4549
364	Angola Basin South Atlantic Ocean	11°34′S 11°58′E	2448

Sapropelic layers are distributed at locations which at the time of sediment accumulation may once have been shallower than 1 km (e.g. Shatsky Rise in the North Pacific, Demerara Rise in the North Atlantic, Walvis Ridge in the South Atlantic, or Mozambique Ridge in the Indian Ocean) as well as in settings that were already as deep as 4 km (North American, Blake Bahama and Cape Basins). Some bathymetric control, however, is observed in that sapropels are only found on the crests of topographic highs in the Pacific Ocean (e.g. Hess, Magellan, Manahiki and Shatsky Rises) and not in time-equivalent strata from the surrounding basins or the flanks of the mid-oceanic ridge. This observation is not evidenced in the North and South Atlantic and also may not hold up for the Indian Ocean. In these latter regions sapropelic horizons are found in the subsurface of abyssal plains that pocketed the center of former deep basins.

A distinction is necessary between (1) those sediments which are rich in organic matter due to preservation by rapid burial of components derived from environments foreign to the deep sea (such as upper-slope carbonaceous terranes where the oxygen-minimum layer impinges on the sea floor or from coastal lagoons, swamps, and foreset deposits off river mouths), and (2) those sediments abundant in plant and animal tissue generated in situ in the ocean and saved from decomposition by the depletion of oxygen at the sediment—water interface.

In the case of the young Mediterranean sapropels, there is a general concensus that the accompanying very thin and delicate laminations illustrated in Fig.2 owe their survival to an annihilation of animal life in and on the substratum and not to sudden entombment. When observed with a hand lens some of the sapropels contain varve-like bedding structures with a frequency that is compatible if each lamina is considered as a window of time of one to a few years duration. No benthonic foraminifera or ostracods are present. Many of the sapropels contain abnormal assemblage of microfossils of planktonic organisms (Olausson, 1965; Ryan, 1972; Cita et al., 1973, 1976). Beds totally devoid of mesopelagic species, in fact, suggest that the poisonous hydrogen-sulfide waters reach upward to levels only a few hundred meters below the sea surface. Beds with intermittent and disproportionately large blooms of phytoplankton such as diatoms and coccoliths indicate the presences of oxygen-depleted water rising well above the shallow thermocline. Because of the broad distribution of piston cores in the

---

Fig.2. Examples of thin millimetric and sub-millimetric thick laminations in eastern Mediterranean sapropels (top) which can be traced in perfect continuity for 500 km between the two cores. Details of the sedimentary layering are preserved by an absence of bottom-dwelling animals during each euxinic phase. Pelagic sediments at times of oxygen ventilation (below) show the effects of bioturbation which obscure the bedding contacts and lithologic boundaries. The core on the left was deposited in a water depth of 2286 m and that on the right at 2858 m. The vertical scale is in centimeters below the top of the core. The excessive thickness between 408 and 412 cm in Core RC9-181 is the result of a thick bed of volcanic ash only thinly represented at 439 cm in Core RC9-185.

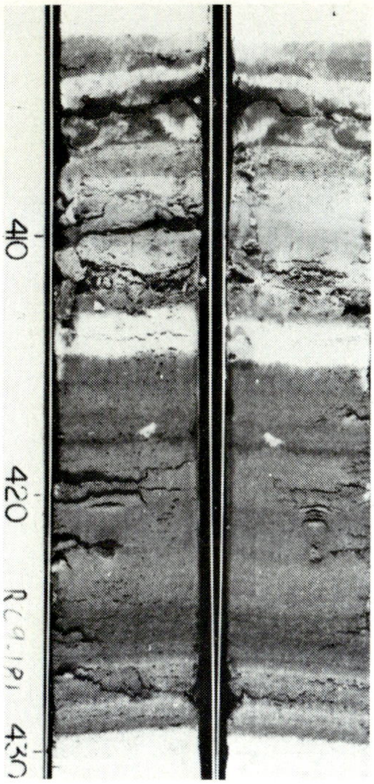
410
420
RC9-181
430

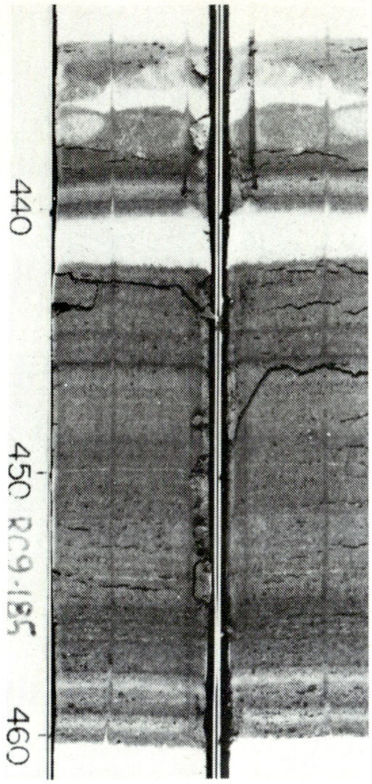
440
450 RC9-185
460

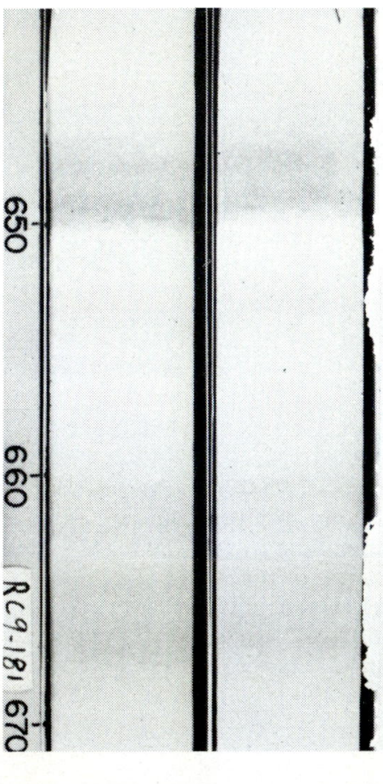
650
660
RC9-181 670

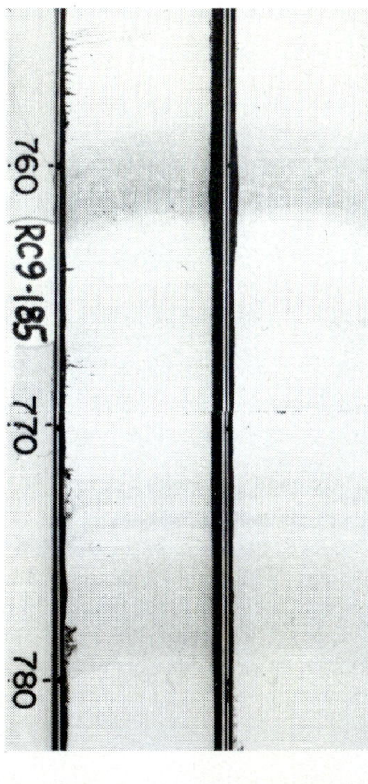
760 RC9-185
770
780

Mediterranean, it has been possible to trace individual sapropels over distances of more than 1000 km (Fig.3) and show that some of these layers begin abruptly on the same millimetric thick lamanae over a water-depth range of several kilometers, extending downwards to the deepest depths and as shallow as 300 m on the continental slope (McCoy, 1974) and across the floor of marginal epeiric seas (Van Straaten, 1972).

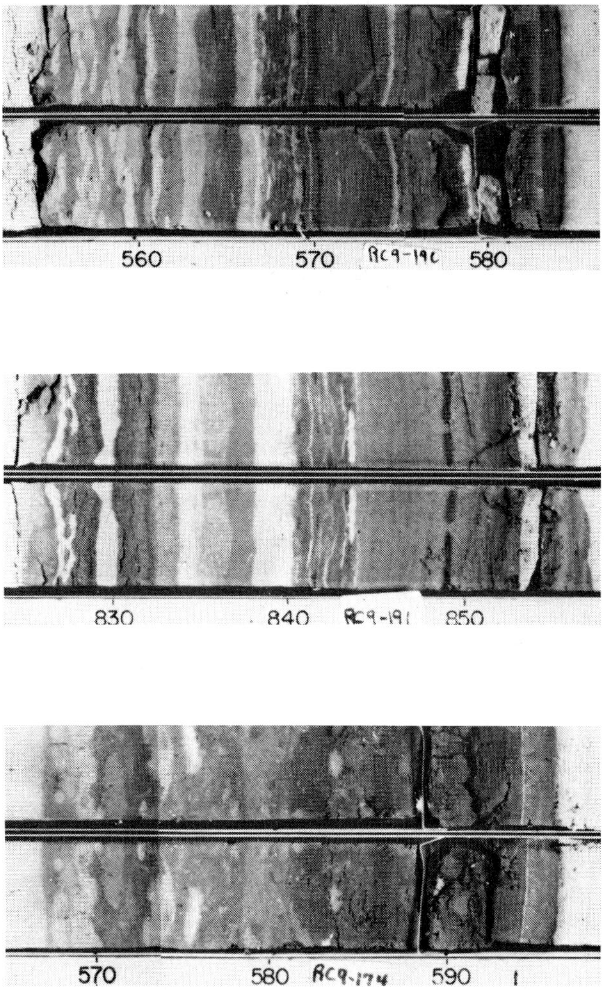

Fig.3. Lateral correlation of the deposit of a single phase of stagnation in the eastern Mediterranean (sapropel S6 of Cita et al., 1976) for a distance of more than 1600 km. The core at the left is from a water depth of 1397 m, the middle from 1712 m, and the right from 2345 m. The thin white layer 2 cm above the base of the sapropel is an ash horizon which is thickest in Core RC9-190 near Italy and thinnest in Core RC9-174 off the coast of Israel. The observed bioturbation in the middle and again near the top of the sapropel results from brief periods of re-ventilation lasting perhaps a few hundred years in duration. The sedimentation rate of these cores averages about 30 m/m.y.

Measurements of the amount of skeletal materials, normalized for sedimentation rate, generally indicate lower fertility of carbonate-secreting organisms during times of sapropel deposition than under conditions of good ventilation. A paucity of calcareous foraminifera is a common situation in the Cretaceous levels. Considerable debate exists as to whether the absence of foraminifera is a consequence of post-burial diagenesis and dissolution (perhaps as a consequence of CO_2 production during the generation of methane by bacteria) or is a reflection of the original absence of these organisms in the Cretaceous seas at the time of the oxygen crises. Some insight into this problem may be gained by noting that a few of the Mediterranean sapropels (admittedly the youngest) still contain thin-walled and translucent pteropods (sapropels S1 and S3 of Cita et al., 1976) and that some of the Atlantic organic shales contain the valves of *Inoceramus* (Site 361) and the delicate shells of ammonites possessing their original nacre (Site 364) (i.e., internal pearly textured aragonitic layer). Many of the Cretaceous sapropels contain calcareous nannoplankton including *Nannoconnus*, which although somewhat less susceptible to dissolution than planktonic foraminifera, nevertheless are usually absent in sediment deposited beneath strongly corrosive bottom waters (Peterson, 1966; Ruddiman and Heezen, 1967; Berger, 1968, 1972). Other Cretaceous sapropels contain only unkeeled forms which are generally thought to be less solution resistant than the keeled ones.

MECHANISMS FOR OXYGEN DEPLETION

The similarity in lithologic makeup, sedimentary structure, thicknesses, and the repetitious nature of many of the carbonaceous Cretaceous horizons from the Atlantic and Caribbean to the much younger Mediterranean sapropels invite a comparison as to the mechanism by which they have formed.

Certain researchers who have investigated the "Glomar Challenger" cores have suggested a vertical expansion of the oxygen-minimum layer on an ocean-wide basis at times of either an augmentation of surface-ocean productivity (analagous to zones of coastal or equatorial up-welling today) or at times of a climatically modulated diminishment in the rate of thermohaline circulation (J. Thiede, personal communication in 1975 following his participation on Leg 39 of the Deep Sea Drilling Project in the South Atlantic; Arthur, 1976; Fischer and Arthur, 1976).

This hypothesis does not require the total removal of oxygen from the seawater and its replacement by hydrogen sulfide, nor does it require a simultaneous euxinification over a broad range of water depths right down to the center of the deepest basins.

For the case of the Pacific Ocean the oxygen-minimum mechanism best explains the distribution of organic-rich sediments which are found on topographic highs and which are the most pronounced near the Cretaceous equator. For the case of the Atlantic there is as yet no clear evidence that indicates that the oxygen removal did not extend right on down to the bottom of the individual basins, as can be definitely documented in the Mediterranean piston cores for the "glacial Pleistocene". Unfortunately, however, there has not yet been adequate drilling over a broad range of paleobathymetric levels for the Cretaceous Atlantic Ocean. Personal experience on Leg 40 in the South Atlantic indicates, however, that those sites at the shallowest depths in the Aptian (for example, Site 363 on the crest of the Walvis Ridge) had significantly fewer, thinner, and less organic sapropels than those near the depositional axis of the basin (Site 361, west of the tip of South Africa).

Unquestionably intervening sills, thresholds and barriers have played an important role. A reconstruction of the Aptian-age water bodies illustrated in Fig.1 (insert) shows the young Indian and South Atlantic Oceans as narrow tracts separated to the east from the large ancestral Pacific by an important structural barrier (the Wallaby Plateaus between Sites 259 and 261), isolated to the southwest from the Pacific by the Falkland Plateau and restricted from the North Atlantic by the equatorial fracture zones between the bulge of South America and identation of West Africa. Furthermore the South Atlantic, itself, is broken into two segments by the important Walvis Ridge and Sao Paulo Plateau escarpments.

In addressing the question of the development of totally euxinic conditions, the stratigraphic resolution for the Cretaceous is not sufficient to demonstrate perfect synchroneity for the sapropel development, since the individual beds are a few tens of centimeters to a few meters in thickness and the precision of correlation is constrained by the width of a faunal zone (generally several millions of years). It is significant, however, that the Late Coniacian events depicted in Fig.1 occur everywhere within the early part of the *Globotruncana concavata concavata* zone (Premoli Silva and Bolli, 1973; and H.M. Bolli and J.F. Longoria, personal communication in 1975).

Stagnation of the Cretaceous North Atlantic would be a straightforward consequence of an ineffective abyssal circulation and a slowdown of convective overturn of these juvenal and somewhat narrow seaways. Depletion of oxygen from bottom water is a natural process taking place in today's ocean and is one of the main mechanisms for keeping a geochemical balance between the carbon sent to the sea floor from dead organisms and that eventually returned to the atmosphere by respiration and decay (i.e. oxidation) at or near the sediment—water interface.

Water sinking to the abyss from the sea surface starts with a dissolved oxygen content in equilibrium with the adjacent atmosphere. Once at depth, however, the oxygen consumed by animals and bacteria can no longer be replaced by atmospheric exchange or plant growth. Today, bottom water spreading into the North Pacific began its journey at the surface of the southern hemisphere polar ocean carrying 8 ml/l of dissolved oxygen. It arrives thousands of miles later with only 2 ml/l, the depletion of 6 ml/l being a measure of significant consumption along route.

If we consider that atmospheric equilibrium in source regions for bottom water in the much warmer Cretaceous oceans would only be about 6 ml/l it is not surprising that even with the vigor of the present Pacific Ocean deep circulation, oxygen starvation is a likely phenomenon. The geological record of ancient stagnation near the abyssal sea floor is thus, in a certain sense, an indicator of the degree of climatic contrast between the polar and equatorial seas.

The sedimentary products of stagnation are also a measure of the magnitude of former sea-surface productivity. For example, life in the surface ocean thoroughly extracts all available nitrogen and phosphorous. These elements are in fact called the biolimiting elements (Broecker, 1974) because the steady-state abundance of living organism depends directly on their availability. In non-stagnant, well-ventilated oceans these elements recycle themselves by way of a more or less continuous vertical mixing, constantly furnishing new opportunities for life, especially in areas of up-welling.

In stagnant oceans, such as today's Black Sea or the Pleistocene Mediterranean, nitrogen becomes incorporated into the sediment reservoir along with carbon in a ratio of ~15:1 close to that which exists in dissolved seawater, suggesting that nitrogen fixation is almost total. As a consequence, it becomes possible to use the nitrogen and carbon contents of buried sapropels and their sediment accumulation rates to calculate past fertility of the organisms now entombed in the sediment. Previously, of course, it was only possible to do this for the organism which left fossilized tests. The productivity rate is in turn an indication of the speed of convective turnover, since it is the upward mixing from the deep-water reservoir which supplies the biolimiting phosphorous. In fact one of the factors causing the periodic large blooms of both epi-pelagic and epi-phytic organisms at times of stagnation (often producing thin white laminae in the sapropels) may be the sudden upwelling of phosphorous-saturated deep water induced by unusual wind patterns or other climatic circumstances. Other thin laminae consisting almost entirely of aragonite needles attest to abrupt supplies of dissolved calcium and bicarbonate.

An interesting insight into the circulation mechanisms may be provided by stable isotopes. Grazzini (in Cita et al., 1976) investigated the isotopic composition of tests of planktonic organisms in eastern Mediterranean sapropels and discovered that the tests of surface-dwelling epi-pelagic taxa such as *Globigerinoides ruber* underwent a range of variation of $\delta\,^{18}O$ of $7.5‰$ greater than that of meso-pelagic species such as *Globerotalia inflata*. Either the surface layer of the Mediterranean experienced a large warming relative to the intermediate and deep waters (perhaps up to 7°C) or the surface layer experienced a higher dilution in isotopically light water at the times of sapropel formation (Olausson, 1965). In either case the evidence points to a marked density stratification as the cause of weakened ventilation.

The Mediterranean sapropels occur generally on warming trends during the glacial cycles (Ryan, 1972) accompanied by rising sea level and perhaps an enhanced supply of fresh-water run off from the melting glaciers of Eurasia.

The >6‰ variation observed between certain glacial to interglacial cycles within the Mediterranean sedimentary sequence (Emiliani, 1955; Cita et al., 1976) is so much larger than the 1‰ change recorded in Caribbean and Pacific sediments (Emiliani, 1966; Shackleton and Opdyke, 1973), that most of the difference seems best accountable to dilution mechanism by ^{18}O depleted runoff and precipitation and only part by temperature variations of a reasonable magnitude. Support for the freshening of surface waters is obtained from abnormal accumulations of planktonic species such as *Globigerina eggeri* (= *Neogloboquadrina dutertrei* [d'Orbigny] of some authors) and *Globigerina bulloides* in certain Mediterranean sapropels (Parker, 1958; Olausson, 1961; Ryan, 1972) and rare oogonia of *Characea* in the Cretaceous carbonaceous clays of the North Atlantic (Luterbacher, 1972). Temperature change may nevertheless have played an associated role by inducing a strong thermal hindrance to convective turnover as a consequence of surface waters warming significantly more rapidly at the beginning of an interglacial interval than bottom waters. Sills and tectonic barriers act to insolate as well as isolate the eastern Mediterranean bottom waters from the dynamically convecting Atlantic Ocean.

Application of the Mediterranean situation to the Mesozoic Atlantic is tempting. The early Cretaceous is for example a known time of widespread outpourings of clastic wedges from major deltas on North America (represented in the very thick Missisagua Formation on the Nova Scotian shelf and the Grand Banks, reported by Sherwin, 1973 and Jansa and Wade, 1975), on Europe (represented in the Wealden deposits in England and the mainland of Europe, reported by Allen, 1969) on Northwest Africa (represented in the Tan Tan formation in the Tarfaya Basin of Morocco) and on South Africa (represented in the Sundays River Formation of the Cape Province of South Africa, reported by Haughton, 1969 and Rigassi and Dixon, 1972). The deep North Atlantic is thought to have been separated from the Pacific by a submarine ridge (the "Panama" barrier discussed in Saunders et al., 1973) as are the eastern Mediterranean basins disconnected from their western counterparts by the shallow straits of Sicily and Gibraltar.

A hypothetical situation envisions a Cretaceous Atlantic undergoing intermittent episodes of excess river input and rainfall leading to an internal density stratification and the exit of a low-density surface-water wedge into the Pacific. Water returning as an intermediate underflow through the Caribbean (the reverse of the present Mediterranean circulation and analogous to the Black Sea) would be oxygen depleted (by organisms in the Pacific equatorial belt of high fertility) and might thereby enhance the oxygen starvation of the Atlantic. If these periods also coincided with warm (or warming) climates and small global variations in surface-water temperature, then bottom-water production by evaporative-generated haline convection would be expected to be minimal and thermal gradients should be weak.

Stagnation of euxinic bottom waters would not necessarily be expected to persist for long intervals due to geothermal heat exiting the ocean floor and thermal conduction and diffusion across water-mass boundaries. Nor would the upper surface of the resulting hydrogen sulfide layer be uniform in depth, being depressed by local evaporative sinking seaward of coastal embayments and around the margins of broad carbonate platforms (e.g. Gulf of Mexico and Blake-Bahama tracts) and elevated beneath belts of high fertility. Density stratification might be most extreme at latitudes of strong precipitation and weakest in areas of arid climate. Hydrogen-sulfide bearing waters might actually have penetrated shallow offshore regions marginal to large coastal deltas, thereby intercalating organic-rich strata into inner-shelf and neritic deposits (such as those examined by the author in southern Morocco and those of the La Luna Formation in the Maracaibo Basin of Venezuela). Upward surges of these poisonous waters may have been responsible for the death and permanent drowning of parts of long linear reefs along the outer margin of North America in the late Neocomian as supported by lithostratigraphic data provided by commercial drilling off Canada (McIver, 1972), by seismic reflection profiling, dredging and coring on the Blake Escarpment (Heezen and Sheridan, 1966) and by DSDP drilling on the Campache Escarpment (Worzel and Bryant et al., 1973).

IMPLICATIONS OF THE OXYGEN CRISES

The organic-matter bearing sediments, deposited in the absences or near absences of oxygen become important geochemical sinks for elemental carbon and sulfur. Carbon is preserved as fossilized animal and plant debris and often includes coaly particles of terrestrial origin (lignite). Abundant amongst the amorphous sapropelic matter are cellular algae, fish scales, spores, pollen grains, marine dinoflagellate cysts and acritarchs (Habib, 1968; 1972; Rossignol-Strick, 1973). Sulfur is retained in the many crystal habits of pyrite and marcasite, and occasionally as nodules and veins of barite. The presence of sulfur becomes most conspicuous when the drill cores are split and allowed to airdry, thereby oxidizing and growing a surficial crust of needle-shaped gypsum crystals.

Carbon is incorporated by planktonic organisms to construct their carbonate hardparts as well as their living tissue. Approximately for every atom of carbon subtracted from atmosphere and oceanic CO_2 and HCO_3^- as skeletal calcium carbonate, there are two atoms removed in organic matter. Most of the organic substances recycle, however, back into the atmosphere and ocean when the plants and animals die and their flesh and fiber oxidize by natural processes of decay and consumption.

Under euxinic conditions in the Cretaceous Atlantic and Indian Oceans decay processes would have been greatly hindered and are expected to have ceased altogether in the deep abyss. By analogy to earlier calculations for the Mediterranean, it is estimated that $80 \cdot 10^{12}$ g/year of carbon might have been permanently locked into the deep-ocean sediment reservoir*.

---

*Using an area of $50 \cdot 10^6$ km^2 for the euxinic sea floor, a mean sediment accumulation rate of 20 m/m.y. and an organic carbon content of 4% in the sapropels.

Considering that the cumulative time of Cretaceous stagnation was at least as great as 1 m.y. (see Fig.1), the net carbon extraction could have exceeded $80 \cdot 10^{18}$ g, a value more than one order of magnitude greater than the identified world reserves for coal and other liquid hydrocarbons, making the Cretaceous a major "carboniferous" period in the history of our planet.

Similar calculations based on an average pyrite concentration of 6% in the sapropels indicate a sulfur extraction of $60 \cdot 10^{18}$ g.

The principal implication of these storages of carbon and sulfur is an increase in the mass of oxygen in the atmosphere. The extent of the increase can be judged by comparing that at present there are $0.7 \cdot 10^{18}$ g of carbon in the atmosphere (as CO_2) and $40 \cdot 10^{18}$ g in the ocean (as CO_2 and HCO_3^-). The Cretaceous net storage of $80 \cdot 10^{18}$ g of carbon corresponds to an atmospheric oxygen increase of 20%. In a like context the net storage of $60 \cdot 10^{18}$ g of sulfur (removed from oceanic SO_4^{2-}) corresponds to an atmospheric oxygen increase of an additional 10%.

As a consequence of the carbon extraction from the oceanic reservoir there should be an enhancement of dissolved calcium in respect to bicarbonate and hence a lowering of the alkalinity of the seawater. Sapropel formation has the effect of denying marine organisms sufficient material for massive shell building and results in widespread dissolution of those tests which are created and then sink towards the deep sea floor.

Hence it is not a coincidence that those intervals of high organic carbon in the DSDP cores, Fig.1, have low contents of calcium carbonate and that those periods rich in limestone, chalk and ooze such as the early Barremian, the Albian and the Santonian—Companian are devoid of extensive bituminous deposits due to effective deep-sea ventilation.

The sulfur extraction is an additional curiosity. Sulfate today is not being supplied to the oceans at a fast rate compared to calcium bicarbonate, but neither is sulfate being used up very rapidly by becoming incorporated into oxidized marine sediments. Hence sulfate stays around in large quantities as the second most abundant anion next to chlorine and three times more abundant than the cation calcium.

One of the major subtracting agents of dissolved sulfate from the ocean is chemical precipitation during the formation of marine evaporites. In the early stages of brine concentration the sulfate ion is lost along with calcium in the crystallization of gypsum or anhydrite. As the calcium is depleted the remaining sulfate becomes available to combine with magnesium. The abundance of the latter ion is sufficient to strip out the rest of the sulfate so that the late-stage salts are generally sodium, magnesium, and potassium chlorides.

During the late Aptian, synchronous with one of the major euxinic phases, an extensive body of salt and evaporite ($>1 \cdot 10^6$ km^3) was laid down in the narrow South Atlantic north of the Walvis Ridge (Belmonte et al., 1965; Brognon and Verrier, 1966; Baumgartner and van Andel, 1971; Leyden et al., 1972). The composition of the deposit is interesting in that beds of $CaCO_3$ and $CaSO_4$ which normally underlie salt sequences of marine origin are lacking at the base of the Brazilian and African deposits. In addition tachyhydrite, a rare $CaCl_2$-bearing evaporite mineral occurs abundantly (Wardlow and Nicholls, 1972), comprising in some deep bore holes up to 15% of the entire sequence.

It is inviting to attribute the sulfate depletion and calcium enrichment observed in the Aptian salt body to the global depletion of sulfur and carbon previously discussed, recalling that the supplying ocean to the south of the Walvis Ridge (Site 361) was strongly euxinic throughout the entire Aptian.

THE BLACK SHALE PROBLEM

Thick sequences of dark-colored organic-matter bearing shales were deposited along many of the continental margins of the Atlantic and Indian Ocean during the early Cretaceous (e.g. Sites 101, 105, 361, 259 and 261). The shales extend out into the subsurface of the modern abyssal plains and are occasionally intercalated with graded layers of sand and silt, exemplified in the massive clastic strata detected in the Cape Basin at Site 361.

The shales are comprised predominantly of clay minerals of terrestrial origin thought to have been carried into the offshore region by submarine currents. They are considered as the distal members of nearshore deltaic sequences explored by commercial drilling. The black shale owes its relatively high organic matter to material transported from the land in rivers, or derived from coastal lagoons and swamps. The dark-colored layers in the deep-sea drill cores contain a conspicuous amount of detrital plant and wood fragments allochthonous to the open-ocean environment in which they were deposited. Although the preservation of the organic matter may have been enhanced by the intermittent oxygen crises, the input of the terrigenous matter could have been entirely independent of local euxinic or non-euxinic conditions. In fact many levels in the Albian and Cenomanian cores from Sites 105, 259 and 261 have been bioturbated by organisms which lived penecontemporaneous with the accumulation of the black shale.

The terrigenous black shales should not be confused with the pelagic sapropels deposited as a direct consequence of deep-ocean oxygen deficiency. Terrigenous black shales may be interbedded by sapropels as at Site 361 or by normal oxidized pelagic marls at Site 261.

The ubiquitous occurrence of black shale in many of the more recent, and therefore unpublished, deep-sea bore holes exclusively within the Cretaceous part of the stratigraphic column has invited some link with the stagnation events. One possible interrelationship might be meteorological changes enhancing global precipitation and leading to the sychronous development of large deltaic prisms and strong density stratification in the ocean surface inhibiting convective overturn in the young narrow Atlantic and Indian Oceans. In this case the Cretaceous stagnations were climatically modulated, as were the Mediterranean ones during the "glacial" Pleistocene.

On the other hand the degree of ancient ventilation of the deep ocean may have had a much greater dependency on physical barriers, including fracture zones (Walvis Ridge), marginal escarpments (Falkland and Wallaby Plateaus), and rising cordilleras (hypothetical "Panama" barrier). In this consideration oxygen starvation may have been a consequence of inefficient bottom-water circulation in numerous isolated basins because of small horizontal thermal gradients. The correlation of euxinic phases with the outbuilding of river deltas may conceivably have been the consequence of partial atmospheric depletion in CO_2 and the resulting decrease in heat retention of the atmosphere. Perhaps there is no interconnection between the two phenomena of black shale and sapropel.

The understanding of cause-and-effect relationships brought about by ocean-wide stagnation is, nevertheless, a serious matter. Many of the aspects of excessive carbon extraction and corresponding perturbations in ocean alkalinity are the exact reverse of those which are enacted today by the burning of fossil fuels. The rates at which these processes occurred may be different, but the magnitude of alteration to the earth's atmosphere and hydrosphere is similar. Hopefully, important lessons may be learned before irreparable change is done to our global environment through our brazen consumption of fossil energy.

ACKNOWLEDGEMENTS

Financial support has been provided by the Division of Ocean Sciences of the U.S. National Science Foundation, Grant No. NSF-OCE-76-02037; the U.S. Department of Commerce, National Oceanic and Atmospheric Administration, Grant No. 03-6-022035120, and Consiglio Nazionale delle Ricerche of Italy (CNR), Comitato 05. The Deep Sea Drilling Project cores were collected aboard the D/V "Glomar Challenger" as part of the National Science Foundation's Ocean Sediment Coring Program by means of a contract with the University of California, Scripps Institution of Oceanography.

Discussions with Michael Arthur, Hans Bolli, Pierre Biscaye, Wallace Broecker, Colette Grazzini, Floyd McCoy, James Natland, Vladimir Nesteroff and Isabella Premoli Silva have been most helpful and are greatly acknowledged.

Special appreciation is expressed to Bruce C. Heezen, convener of the symposium, for introducing W.B.F.R. to the problem of sapropels and sharing his many thoughts on the effects of deep-ocean circulation in time and space.

REFERENCES

Allen, P., 1969. Lower Cretaceous sourcelands and the North Atlantic. Nature, 222: 657—658.

Andrews, J.E. and Packham, G. et al., 1975. Initial Reports of the Deep Sea Drilling Project, 30. U.S. Government Printing Office, Washington, D.C., pp. 1—753.

Arthur, M.A., 1976. The oxygen minimum: expansion, intensification, and relation to climate, Abstract. Joint Oceanographic Assembly, Edinburgh.

Baumgartner, T.R. and van Andel, T.H., 1971. Diapirs of the continental margin of Angola. Geol. Soc. Am. Bull., 82: 793—802.

Belmonte, Y., Hirtz, P. and Wenger, R., 1965. The salt basins of the Gabon and the Congo (Brazzaville), a tentative paleogeographic interpretation. In: Salt Basins around Africa. London Inst. Petroleum, London, pp.55—74.

Berger, W.H., 1968. Planktonic Foraminifera: selective solution and paleoclimatic interpretation. Deep-Sea Res., 15: 31—43.

Berger, W.H., 1972. Deep-sea carbonates: dissolution facies and age—depth constancy. Nature, 236: 392—395.

Berger, W.H. and von Rad, U., 1972. Cretaceous and Cenozoic sediments from the Atlantic Ocean. In: D.E. Hayes and A.C. Pimm et al., Initial Reports of the Deep Sea Drilling Project, 14. U.S. Government Printing Office, Washington, D.C., pp.784—954.

Broecker, W.S., 1974.Chemical Oceanography.Harcourt Brace Jovanovich, New York, N.Y., pp.1—214.

Brogon, G.P. and Verrier, G.R., 1966. Oil and geology in Cuanza basin of Angola. Bull. Am. Assoc. Pet. Geol., 50: 108—158.

Bukry, D., Kling, S.A., Horn, M.K. and Manheim, F.T., 1970. Geological significance of coccoliths in fine-grained post-glacial carbonate bands of the Black Sea sediments. Nature, 226: 156—158.

Cita, M.B., Chierici, M.A., Ciampo, G., Moncharmont Zei, M., d'Onofrio, S., Ryan, W.B.F. and Scoziello, R., 1973. The Quaternary record in the Tyrrhenian and Ionian Basins of the Mediterranean Sea. In: W.B.F. Ryan and K.J. Hsu et al., Initial Reports of the Deep Sea Drilling Project, 13. U.S. Government Printing Office, Washington, D.C., pp.1263—1339.

Cita, M.B., Grazzini, C., Chamley, H., Ciaranfi, N., d'Onofrio, S. and Robert, C., 1976. Paleontological, isotopic, mineralogical and sedimentological investigations on a deep-sea core from the Mediterranean Ridge in the Ionian Basin. Quat. Res., in press.

Degens, E.T. and Ross, D.A. (Editors), 1973. The Black Sea: Its Geology, Chemistry and Biology. Mem. Am. Assoc. Pet. Geol., 20: 1—663.

Deuser, W.G., 1973. Evolution of anoxic conditions in the Black Sea during the Holocene. In: E.T. Degen and D.A. Ross (Editors), The Black Sea: Its Geology, Chemistry and Biology. Mem. Am. Assoc. Pet. Geol., 20: 133—136.

Emiliani, C., 1955. Pleistocene temperature variations in the Mediterranean. Quaternaria, 2: 87—98.

Emiliani, C., 1966. Paleotemperature analysis of the Caribbean core P6304-8 and P6304-9 and a generalized temperature curve for the last 425,000 years. J. Geol., 74: 109—126.

Fischer, A.G. and Arthur, M.A., 1976. Secular variations in the pelagic realms. In: H.E. Cook and P. Enos (Editors), Basinal Carbonate Sediments. Soc. Econ. Paleontol. Mineral. Spec. Publ., in press.

Foresman, J.B., 1976. Organic Geochemistry. In: H.M. Bolli and W.B.F. Ryan et al., Initial Reports of the Deep Sea Drilling Project, 40. U.S. Government Printing Office, Washington, D.C., in press.

Girdley, W.A., Leclaire, L., Moore, C., Vallier, T.L. and White, S.M., 1974. Lithologic summary, Leg 25, Deep Sea Drilling Project. In: E.S.W. Simpson and R. Schlich et al., Initial Reports of the Deep Sea Drilling Project, 25. U.S. Government Printing Office, Washington, D.C., pp.725—741.

Habib, D., 1968. Spores, pollen, and microplankton from Horizon Beta outcrop. Science, 162: 1480—1485.

Habib, D., 1972. Dinoflagellate stratigraphy, Leg 11, Deep Sea Drilling Project. In: C.D. Hollister and J.I. Ewing et al., Initial Reports of the Deep Sea Drilling Project, 11. U.S. Government Printing Office, Washington, D.C., pp.367—381.

Haughton, S.H., 1969. Geological History of Southern Africa. Geological Society of South Africa Cape and Transvaal Printers, Cape Town, pp. 1—535.

Heezen, B.C. and Sheridan, R.E., 1966. Lower Cretaceous rocks (Neocomian—Albian) dredged from Blake Escarpment. Science, 154: 1644—1647.

Jansa, L.F. and Wade, J.A., 1975. Geology of the continental margin off Nova Scotia and Newfoundland. In: Offshore Geology of Eastern Canada. Geol. Surv. Can. Pap., 74-30, 2: 51—105.

Kidd, R.B., Cita, M.B., McCoy, F.H. and Ryan, W.B.F., 1976. The stratigraphy of eastern Mediterranean sapropel sequences recovered by DSDP Leg 42A and their paleoenvironmental significance. In: K.J. Hsu and L. Montadert et al., Initial Reports of the Deep Sea Drilling Project, 42A. U.S. Government Printing Office, Washington, D.C., in press.

Lancelot, Y., Hathaway, J.C. and Hollister, C.D., 1972. Lithology of sediments from the western North Atlantic, Leg 11, Deep Sea Drilling Project. In: C.D. Hollister and J.I. Ewing et al., Initial Reports of the Deep Sea Drilling Project, 11. U.S. Government Printing Office, Washington, D.C., pp.901—949.

Larson, R.L. and Moberly, R. et al., 1975. Initial Reports of the Deep Sea Drilling Project, 32. U.S. Government Printing Office, Washington, D.C., pp. 75—158.

Leyden, R., Bryan, G. and Ewing, M., 1972. Geophysical reconaissance on African shelf, 2. Margin sediments from Gulf of Guinea to Walvis Ridge. Bull. Am. Assoc. Pet. Geol., 56: 682—693.

Lisitzin, A.P., 1972. Sedimentation in the World Ocean. (Edited by K.S. Rodolfo). Soc. Econ. Paleontol. Mineral. Spec. Publ., 17: 1—218.

Luterbacher, H., 1972. Foraminifera from the Lower Cretaceous and Upper Jurassic of the Northwestern Atlantic. In: C.D. Hollister and J.I. Ewing et al., Initial Reports of the Deep Sea Drilling Project, 11. U.S. Government Printing Office, Washington, D.C., pp.561—593.

McIver, N.L., 1972. Cenozoic and Mesozoic stratigraphy of the Nova Scotia Shelf. Can. Jour. Earth Sci., 9: 54—70.

McCoy Jr., F.W., 1974. Late Quaternary Sedimentation in the Eastern Mediterranean Sea. Thesis, Harvard University, Cambridge, Mass., pp.1—132 (unpublished).

Menard, H.W. and Smith, S.M., 1966. Hypsometry of ocean basin provinces. J. Geophys. Res., 71: 4305—4325.

Nesteroff, W.D., 1973. Petrography and mineralogy of sapropels. In: W.B.F. Ryan and K.J. Hsu et al., Initial Reports of the Deep Sea Drilling Project, 13. U.S. Government Printing Office, Washington, D.C., pp.713—720.

Olausson, E., 1961. Studies of deep-sea cores. Rep. Swed. Deep-Sea Exped., 1947—48, 8: 353—391.

Olausson, E., 1965. Evidence of climatic changes in North Atalntic deep-sea cores. Progress in Oceanography, Pergamon Press, New York, 3: 221—254.

Parker, F.L., 1958. Eastern Mediterranean Foraminifera. Rept. Swed. Deep-Sea Exped., 1947—48, 8: 219—283.

Peterson, M.N.A., 1966. Calcite: rates of dissolution in a vertical profile in the central Pacific. Science, 154: 1542—1544.

Premoli Silva, I. and Bolli, H.M., 1973. Late Cretaceous to Eocene planktonic foraminifera and stratigraphy of Leg 15 sites in the Caribbean Sea. In: N.T. Edgar and J.B. Saunders et al., Initial Reports of the Deep Sea Drilling Project, 15, U.S. Government Printing Office, Washington, D.C., pp. 449—547.

Rigassi, D.A. and Dixon, G.E., 1972. Cretaceous of the Cape Province, Republic of South Africa. Proc. Ibadan Univ. Conf. on African Geology, 1970, Ibadan, Nigeria, pp.513—527.

Robinson, P.T., Thayer, P.A., Cook, P.J. and McKnight, B.K., 1974. Lithology of Mesozoic and Cenozoic sediments of the eastern Indian Ocean, Leg 27, Deep Sea Drilling Project. In: J.J. Veevers and J.R. Heirtzler et al., Initial Reports of the Deep Sea Drilling Project, 27. U.S. Government Printing Office, Washington, D.C., pp.1001—1047.

Ross, D.A., Degens, E.T. and MacIlvaine, J., 1970. Black Sea: recent sedimentary history. Science, 170: 163—165.

Rossignol-Strick, M., 1973. Pollen analyses of some sapropel layers from the deep-sea

floor of the eastern Mediterranean. In: W.B.F. Ryan and K.J. Hsu et al., Initial Reports of the Deep Sea Drilling Project, 13. U.S. Government Printing Office, Washington, D.C., pp.971—991.
Ruddiman, W.F. and Heezen, B.C., 1967. Differential solution of planktonic Foraminifera. Deep-Sea Res., 14: 801—808.
Ryan, W.B.F., 1972. Stratigraphy of Late Quaternary sediments in the eastern Mediterranean. In: D.J. Stanley (Editor), The Mediterranean Sea. Dowden, Hutchinson and Ross, Stroudsburg, Pa., pp.149—169.
Saunders, J.B., Edgar, N.T., Donnelly, T.W. and Hay, W.W., 1973. Cruise synthesis. In: N.T. Edgar and J.B. Saunders et al., Initial Reports of the Deep Sea Drilling Project, 15. U.S. Government Printing Office, Washington, D.C., pp.1077—1111.
Shackleton, N.J. and Opdyke, N.D., 1973. Oxygen isotope and paleomagnetic stratigraphy of equatorial Pacific core V28-238: oxygen isotope temperature and ice volumes on a 10^5 year and 10^6 year scale. J. Quat. Res., 3: 39—55.
Sherwin, D.F., 1973. Scotian Shelf and Grand Banks. In: R.G. McCrossan (Editor), Future Petroleum Provinces of Canada — Their Geology and Potential. Can. Soc. Pet. Geol. Mem., 1: 519—559.
Van Hinte, J.E., 1976. A Cretaceous time scale. Bull. Am. Assoc. Pet. Geol., 60: 498—516.
Van Straaten, L.M.J.V., 1972. Holocene stages of oxygen depletion in deep waters of the Adriatic Sea. In: D.J. Stanley (Editor), The Mediterranean Sea. Dowden, Hutchinson and Ross, Stroudsburg, Pa., pp.631—643.
Wardlaw, N.C. and Nicholls, G.D., 1972. Cretaceous evaporites of Brazil and West Africa and their bearing on the theory of continental separation. Proc. Int. Geol. Congr., 24 Sess., Sect. 6: 43—55.
Worzel, J.L., Bryant, W. et al., 1973. Initial Reports of the Deep Sea Drilling Project, 10. U.S. Government Printing Office, Washington, D.C., pp.25—47.